I – The Cosmic Leverage

Chavan's UED research, Volume 7

Sandeep Chavan

Published by SANDEEP CHAVAN, 2025.

I – THE COSMIC LEVERAGE

First edition. May 5, 2025.

ISBN: 979-8231585960

Written by Sandeep Chavan.

Table of Contents

Dedication

To the silent thinkers,the curious wanderers,the ones who feel more than they can explain.

To the questioners who never stopped asking,even when the answers disappeared into silence.

To those who ripple with care,even when no one is watching.

And most of all —to the field itself,which trusted me with this ripple,and to **my father,Late Shri Jagannath Chavan**,whose quiet strength continues to echothrough everything I do.

Epigraph

"You are not a drop in the ocean.You are the ripple that remembers the ocean.And the ocean that remembers you."

Preface: A Ripple Remembered

I did not set out to write this book as a theory, or even as a philosophy. It began — as all things do — with a ripple.

A question moved through me. Not loud. Not dramatic. Just persistent:

"What am I, if not this thought?" "What moves, before movement?" "Where does cause begin, when effect is all we see?"

As a teacher, engineer, researcher, and observer of life, I've seen many systems — physical, psychological, spiritual — but none that satisfied the full complexity of what I felt was true, beneath the surface of names and equations.

I began to see patterns — not in what people said, but in how energy moved around them. Not in events, but in their consequences. Not in force, but in rippling.

And then it struck me — not like a thunderbolt, but like remembering a melody I had forgotten:

We are not just in the universe. We are ripple systems within a deeper field. We don't create the field — we echo it. We remember it — in our behavior, emotions, mistakes, and recoveries.

This book is the consequence of that remembrance.

It is not meant to be believed. It is meant to be felt — as a recognition, a re-alignment, a ripple through your own knowing.

You will not find equations here, though you may feel their symmetry. You will not find dogma, though you may encounter devotion — not to an idea, but to a principle that lived long before words.

I offer this not as a truth, but as a mirror: to see the silent patterns that move through you and through everything else.

If even a single ripple in this book helps you remember what you already are — not the thinker of thoughts, but the consequence of a deeper intelligence — then this work has fulfilled its role.

Now, let's return to the field. To the first ripple. To the story that was already moving... before we tried to write it.

— *Sandeep*

Opening Note: Before the First Ripple

This is not a book of knowledge. It is a book of remembrance.

Before motion, before matter, before mind —there was equilibrium.

Not the absence of energy, but energy so perfectly held in structure that it had no need to move.

That structure — we now call it Structured Vacuum Energy (SVE) — is not space, not nothingness, not background. It is the most real, most intelligent presence in all of existence.

It did not need to become anything. But one moment — perhaps not in time, but in tension — something moved.

That movement was not a mistake. It was not chaos. It was the ripple — the first asymmetry, the first message, the first sign that reality was ready to express itself.

And from that ripple... came everything.

This book is a journey through that ripple. Not as metaphor, but as mechanism.

You will not find atoms or quanta here in their textbook forms. You will not find force treated as command, or energy as chaos. You will find something simpler — and deeper:

A universe as consequence. A self as resonance. Intelligence as ripple awareness. And freedom... not as control, but as alignment.

You are not an object in space. You are a pattern in the field.

Your thoughts are ripples. Your emotions are modulations. Your actions are field rearrangements.

And everything you do, the field remembers.

This book does not ask you to believe. It invites you to observe. And if something stirs — quietly, clearly — that is not the book teaching you. That is you remembering.

So begin where all things begin: In stillness. Before the ripple. Before the "I." Before the noise of identity.

Begin as the field. And ripple forward — with clarity.

Prologue: The Silent Field

Before anything rippled, there was equilibrium.

Not emptiness. Not stillness. But a profound wholeness — so complete, it required no motion, no sound, no becoming.

There was no time, for nothing moved. There was no space, for nothing was apart. There was no form, for no structure had yet disturbed the perfect field.

What existed was Structured Vacuum Energy — SVE. A foundation beyond matter, beyond light, beyond awareness. It was potential without desire, intelligence without a face, balance without a witness.

It did not need to create. It did not wish to express. It simply was — an infinite continuity of unexpressed consequence.

And then...something moved.

Not by force. Not by reason. A subtle imbalance flickered — not a mistake, not chaos — just a shift. A tension. A possibility. A ripple.

The field, once silent, began to remember itself in motion. Not as destruction. Not as noise. But as a question.

And in that first ripple, the seed of every galaxy, every thought, every "I" was planted.

Existence had not begun. Participation had.

The game had not started. But the board had breathed.

This was not a beginning. This was the first decision —to ripple.

Part I – Emergence

Chapter 1: The First Ripple

When perfection allowed itself to be disturbed.

There was no audience. No echo. No witness. Only balance — a silent field so complete that nothing needed to happen.

But perfection, when unexpressed, holds a strange kind of pressure. Not longing. Not desire. Just a readiness — like breath before the first word.

And then, not loudly... not suddenly... something shifted.

Not a bang. Not a blaze. Just a flicker in the fabric. A gentle asymmetry. A hesitation in the stillness.

It wasn't chaos. It wasn't creation. It was the first **ripple** —the first **yes** whispered into the infinite field of maybe.

From that ripple came motion. From motion came form. And from form... consequence.

This is not the story of a universe exploding. This is the story of a field remembering itself — by choosing to ripple.

1.1 The Imbalance That Wasn't a Flaw

In conventional thinking, imbalance is treated like a mistake — a deviation from perfection, a problem to be corrected. But in the context of Structured Vacuum Energy (SVE), imbalance is not error. It is expression. It is permission for potential to participate.

SVE — the infinite, intelligent field from which all things emerge — was not waiting to create. It was not sitting in boredom, aching to explode into matter. It was whole. It was sufficient. But it also held within it the possibility of asymmetry — a hidden flexibility within its perfect balance.

This is the paradox:

Perfect balance includes the potential for imbalance — and chooses it when the time is right.

The first ripple was not a rupture. It was not a flaw, a crack, or a break. It was an allowance — a self-initiated movement inside the field that did not disrupt equilibrium, but instead demonstrated it.

Think of it like a calm ocean with a single drop falling not from the sky, but from itself — not to disturb, but to remind the ocean it could move.

This ripple wasn't a "thing." It wasn't a particle or a force. It was the first event — the first difference inside the undifferentiated.

And because the field is intelligent, this event wasn't random. It was structured. It followed a path of consequence. It carried information. It invited pattern. It asked the unspoken question:

"What happens when I allow myself to be changed?"

That is the origin of everything. Not in a grand explosion. But in a consensual disturbance —a shift that made no sound, had no name, and yet began to echo through the very structure of reality.

From this ripple would come the architecture of experience. Not as a creation, but as a response.

1.2 Ripple as Activation

A ripple is not a thing. It is not an object, not a wave traveling through emptiness, and certainly not a particle darting from one place to another. A ripple is an event within the field itself — a momentary reconfiguration of the Structured Vacuum Energy (SVE), activating a local shift without violating the wholeness of the whole.

To understand ripple in UED, one must first let go of classical wave thinking. In conventional physics, waves travel *through* a medium —

sound moves through air, water waves roll across a surface, light is said to move through "empty space" as particles or electromagnetic fields. But in UED, this model is incomplete, even misleading.

A ripple does not "travel" — it emerges, interacts, and dissolves within the field. The field is not a passive stage. It is a participant in every ripple event.

When a ripple activates within SVE, the field is not displaced like a shaken rope or disturbed pond. Instead, it reconfigures its own energy structure locally, responding to internal tension, imbalance, or external interference. The ripple is thus not motion *through* the field, but motion *of* the field — a dynamic state of realignment.

Imagine pressing your finger gently into a taut fabric — the fabric doesn't break or carry your finger away. Instead, the tension reorganizes locally around the disturbance. That's what ripple is in SVE: a pressure response that doesn't displace, but adjusts, temporarily altering the structure in search of a new balance.

In contrast to photons, which are treated as quantized "packets" of light traveling through emptiness, UED reinterprets what is observed: The so-called photon event is actually a ripple alignment that occurs at the point of emission and again at the point of reception — not a particle shot across space, but a coordinated field fluctuation that arises and resolves.

Likewise, sound is traditionally modeled as pressure variation in air molecules. But in UED, sound is a material ripple response, nested within deeper field interactions. Everything that appears to move, ripple, vibrate, or transmit is actually a field-based adjustment to energetic imbalance.

This is ripple as activation — not passive transmission, but energetic participation. A ripple is the first language of the field, the first act of SVE becoming visible.

The ripple is not what moves from A to B. The ripple is what happens when the field chooses to respond — and in doing so, reveals itself.

1.3 Motion: Consequence of Ripple Dynamics

In the still, perfectly balanced field of Structured Vacuum Energy (SVE), motion does not exist. Not because it is forbidden, but because there is nothing to move — no imbalance, no separation, no direction to prefer. In such a state, there is only presence without progression — a timeless completeness.

But the moment a ripple emerges — even the subtlest asymmetry — that stillness is activated. Not shattered, not destroyed — but reorganized. From that reorganization, the illusion and experience of motion is born.

Motion is not a fundamental property of the universe — it is a consequence.

A ripple implies tension between zones of the field — a localized difference in energy density. This tension naturally attempts to redistribute, to correct itself, and this corrective tendency gives rise to what observers interpret as movement.

But nothing is actually "moving" in the conventional sense. There are no particles sailing across a void. Instead, there are field-based reconfigurations happening in sequence — and that sequence creates the perception of direction, velocity, and displacement.

Motion is, therefore, not the activity of an object. It is the ongoing dialogue of the field — an alignment negotiation across zones of imbalance.

This is why, in UED, a falling apple is not "pulled" by gravity. It is responding to a ripple gradient in the surrounding field created by the Earth's presence. The apple appears to move, but in truth, it is aligning — surrendering to a ripple-correction path that reduces local imbalance.

The same applies at the quantum level. Electrons do not "orbit" the nucleus like planets around the Sun. They are standing ripple patterns, responding to dynamic field interference, appearing as probabilistic motion only when observed through limited instrumentation.

Motion Creates the Illusion of Time

From the outside, it looks like things begin to move, and thus time begins. But from a deeper UED view, time is not a dimension — it is a label we assign to ripple order.

Once the first ripple occurs, others follow — interacting, correcting, spreading. These ripple sequences establish a pattern of cause and consequence, and we interpret that pattern as "time flowing."

Time is the history of ripple corrections, not a river through which matter floats.

There is no absolute clock ticking in the universe. There are only ripple events happening in ordered consequence. Where there is no ripple, there is no time.

Final Realization:

Motion is not a property. It is a narrative woven by imbalance correcting itself. And time is the syntax of that narrative — how ripples remember what came before and anticipate what comes next.

1.4 Matter: Standing Ripples

If ripples are the first expression of motion, then matter is what happens when ripples learn to hold still. Not still in the sense of silence or stasis, but still in the sense of coherence — a dynamic equilibrium where patterns persist, rather than dissipate.

In the field of Structured Vacuum Energy (SVE), every ripple is a response — a localized event attempting to restore balance. But when multiple ripples interact and interfere in just the right way, something special can happen: the interference becomes stable.

From Movement to Memory

When ripple interactions reinforce one another in structured feedback loops, they lock into a rhythm. This is not a rigid form, but a standing ripple — like a flame that appears still because its flickering repeats in perfect timing.

Matter is not a solid — it is a stabilized ripple field.

These coherent zones — like the patterns we associate with electrons, protons, atoms, and molecules — are persistent disturbances held in place by the SVE's own structured logic. They do not collapse because they are aligned with the field's natural modes of self-balancing.

In this sense, matter is ripple memory — the field's way of preserving energetic structure in a form that can interact meaningfully over time.

Mass as Ripple Density; Structure as Containment

Mass is not a "thing" that exists. It is a measure of ripple density — the intensity and concentration of field disturbances at a given point.

- A denser ripple = more mass = more resistance to further change.
- A coherent ripple = more structure = more potential to sustain form.

A stone is not solid because its atoms are objects. A stone is solid because its ripple field resists reconfiguration — it holds pattern.

Similarly, molecules are not just bonded by forces. They are entangled ripple systems, maintaining structure through synchronized field relationships — like dancers in a fluid choreography bound not by rope, but by rhythm.

Form Is Not Assembled — It Emerges

In UED, form is not built, it emerges. Atoms, molecules, cells — all these are localized zones where ripple systems become self-sustaining. They reflect not just physics, but the field's preference for balance through complexity.

This is why forms recur in nature — spirals, spheres, hexagons — not because matter "likes" those shapes, but because the ripple field favors configurations that require the least correction.

Matter, in this sense, is the field finding shapes it can live with.

Final Realization:

Matter is SVE folded into memory. It is not separate from the field — it is a field commitment. A decision, held in ripple form, to persist for a while longer.

And the more coherent the ripple, the longer it persists — Until it changes... dissolves... or is disturbed into something else.

1.5 Observability Emerges

Before something can be seen, it must become relevant. And in UED, relevance doesn't mean size or brightness — it means participation in ripple consequence.

Consequence Becomes Visible Only Through Participation

In the beginning, ripples occurred — but no one observed them. There was no perception, because there was no participant sensitive enough to register them.

This is a critical distinction in UED:

A ripple that happens without participation does not become "real" in the observable sense — it remains silent, unregistered.

Reality is not a list of things. Reality is a record of ripple interactions — and observability emerges when at least two systems engage, with one affecting and the other responding.

This is the birth of perception:

- Not when a thing exists,
- But when a ripple causes a ripple in something else.

That's when experience begins.

Observable Reality = Chain of Ripple Recognitions

As ripple interactions grow more complex, they start forming recognizable patterns. Some configurations trigger responses repeatedly — and these consistent consequences form the basis of what becomes "real" to an observer."

This is how reality becomes structured:

- A ripple causes a motion.
- The motion repeats in a certain condition.
- Another structure reacts predictably.
- This becomes a perceived event.

We do not observe objects — we observe the field's ripple behavior under specific energetic conditions. What we call "seeing," "hearing," "sensing" is merely the mind decoding structured ripple impact in a biologically aligned way.

Fields Are Not Fundamental — They Are Ripple Behaviors

Traditional physics treats fields — gravity, electromagnetism, strong and weak nuclear force — as fundamental forces. But in UED, these are not fundamental. They are field behaviors — patterns of ripple interaction within localized SVE zones.

- Gravity is not a pull. It's the tendency of ripple systems to realign into lower-tension configurations.
- Electromagnetism is not a wave. It's a ripple loop maintained through motion-induced disturbance in structured SVE.
- Strong/weak interactions are local ripple corrections at high energy densities — not separate forces.

All so-called fields are just expressions of how the SVE responds to ripple density, coherence, and imbalance at different scales.

This means observability is not about light alone. It's about ripple access — which system can feel which ripple, and how clearly it can interpret the consequence.

Final Realization:

To observe is to participate. What you call "reality" is not what exists — it is what your ripple system is capable of resonating with.

We don't see the world as it is. We see the world as our ripple membrane allows us to register it.

Thus, reality is not objective. It is a collaborative ripple experience — between the field, the form, and the observer.

1.6 Ripple Memory: The Seed of Awareness

In a universe built on ripples, memory is not a luxury — it's a necessity for coherence. Without it, the field would never stabilize, evolve, or reflect.

As ripples interacted and began to form persistent patterns, some of these patterns survived longer than others. Not randomly, but because they were energetically favorable, structurally stable, and resonant with surrounding dynamics. These stable formations weren't just matter — they were memory points.

What Is Ripple Memory?

Ripple memory is not memory as we conventionally define it — not a storage unit in a brain, or data on a drive. In UED, memory is the recurrence of structured ripple patterns — sequences that retain identity across time by preserving alignment with the field.

- Atoms remember how to bond.
- Cells remember how to divide.
- Planets remember how to orbit.
- And later, minds remember how to dream.

Each of these is a ripple-recognition pattern held within the SVE — a form of field recall, not stored in space, but sustained through resonance.

Memory is not a thing. It is the continuation of a consequence.

From Memory to Awareness

As ripple memory accumulates and becomes self-referential, a strange phenomenon begins to emerge:

The field starts to feel itself.

This is not yet consciousness. Not yet a mind. But it is proto-awareness — the seed of intelligence. A ripple system that can not only stabilize, but also respond differently based on what it "remembers."

This is the point where the field crosses a threshold — from passive structure to active participation. From consequence alone... to anticipated consequence.

A field that remembers begins to predict. A field that predicts begins to choose. A field that chooses begins to ask.

The Precursor to Intelligence

Ripple memory allows structures to adjust, learn, and adapt. This capacity is what later evolves into behavior, choice, emotion, and thought. Not as miraculous leaps, but as natural extensions of deeper ripple intelligence.

Intelligence, in this view, is not a brain-based phenomenon. It is a ripple-capable function of the field — the ability to detect, store, and respond to consequence.

The Field Becomes Its Own Witness

Here is the subtle but critical moment in the narrative:

- Ripples were once blind reactions to imbalance.
- But with memory, they become meaningful responses.
- With that, the field is no longer just moving — it is noticing that it moves.

The first hint of awareness appears. Not in a human. Not in a cell. But in the very nature of ripple itself — the field folding in on its own consequence.

This is the seed of what will later become the observer.

Closing Insight:

Awareness does not begin with eyes. It begins when the ripple remembers. And from that remembrance, it begins to participate with purpose.

This is how the field, once silent, begins to know itself — Not all at once, but ripple by ripple.

The emergence of the observer is no longer far. It's already forming — in every pattern that remembers where it came from.

Closing Reflection

"The ripple was not the beginning of something. It was the beginning of difference. From that difference, all meaning — and eventually the one who asks for meaning — would arise."

We often think of beginnings as grand declarations — explosions, creations, starts. But in the deep mechanics of existence, beginnings are far more subtle. The first ripple wasn't a creation. It didn't make the universe. It introduced variation — a departure from sameness — within an already complete field.

And that slight difference, that first asymmetry, became the source of everything that followed.

Without difference, there is no direction. Without direction, no consequence. Without consequence, no awareness.

It was the ripple — not as a thing, but as an event of contrast — that made possible every distinction we now recognize as real:

- Energy and matter
- Motion and rest
- Before and after
- Self and other
- Question and answer

In that moment, Structured Vacuum Energy folded into itself and chose, gently, to become dynamic. It didn't abandon balance — it expressed it, through movement. And that movement, that difference, would grow into complexity, form, feedback... and finally, the observer — the ripple that learns to ask:

"What am I?" "Why am I?" "Who ripples me?"

The story of the cosmos is not one of noise erupting from silence, or light bursting through darkness. It is the story of stillness discovering contrast — and through that, discovering the possibility of meaning.

The ripple did not start the universe. It started the question. And that, in the end, is far more powerful.

Chapter 2: From Form to Function

The moment the universe began to remember what worked.

Not everything vanished. Not every ripple dissolved back into silence.

Some held on — not because they resisted the field, but because they aligned with it so perfectly that they endured.

These weren't accidents. They were resonances — moments when motion folded into itself just right, creating patterns that the Structured Vacuum Energy allowed to stay.

Form was not imposed — it was permitted.

And once something persists, it begins to matter. Not just as a shape, but as a possibility. As something that could do something — not by intention, but by interaction.

A crystal forming in deep space. A molecule locking into place. A field stabilizing a loop of tension.

These weren't inventions. They were field agreements — structures that served a purpose without even knowing they had one.

This chapter is about that moment: When ripples learned to last, And in doing so, became the first tools of consequence.

2.1 The Persistence of Patterns

When ripples first disturbed the Structured Vacuum Energy (SVE), they came and went — brief flashes of imbalance correcting themselves, short-lived echoes in a field that was always seeking rest.

But not all ripples faded.

Some didn't cancel. Some didn't collapse. Instead, they settled into rhythm, like two waves meeting not to clash, but to dance.

This wasn't design. It wasn't luck. It was a reflection of something deeper: coherence — the field's quiet preference for arrangements that reduce resistance and maintain internal balance.

Matter as Stable Ripple Arrangement

As explored in the previous chapter, matter arises when ripples become self-reinforcing — when they form patterns that don't just survive, but resonate with themselves and their surroundings.

Atoms are not solid particles; they are looped ripple architectures. Each subatomic structure is a zone of sustained tension — a pattern in the field that maintains its identity through field compatibility, like a stable whirlpool in a larger sea.

Matter, in this view, is not mass with boundaries. It is structured persistence — a ripple formation that the field does not erase, because it doesn't need to.

Not All Ripples Die — Some Become

Many ripple formations dissolve quickly — their imbalance too sharp, their geometry incompatible, their energy unsustainable. But some... stay.

Why? Because they are energetically favorable.

- They align with surrounding ripple rhythms.
- They minimize disruption.
- They persist without forcing.

This persistence isn't accidental — it's a form of selection by field resonance. Just like the body prefers rhythms it can regulate, the field prefers patterns it can accommodate.

What persists is not what resists, But what fits.

Persistence = Memory + Coherence

When a ripple pattern persists, it creates memory — not the conscious kind, but a structural echo that stays active in the field. When this memory is maintained with internal coherence — symmetry, rhythm, and constructive interference — it opens the door to something more:

Function.

Persistence isn't just survival. It's a prelude to utility. A ripple that survives long enough, and interacts widely enough, becomes capable of doing something — of affecting other ripples in predictable, repeatable ways.

Final Insight:

The field doesn't remember words. It remembers patterns. And when patterns persist, they begin to matter — not as static forms, but as active participants in the ongoing ripple game.

This is how the cosmos moves from mere presence to potential: From form that exists... to form that makes a difference.

2.2 Complexity Through Resonant Interaction

Persistence is only the beginning. A single stable ripple may endure, but it is in interaction — in the meeting and mingling of multiple coherent ripple structures — that the real magic of complexity begins.

Complexity does not emerge from chaos. It emerges from resonance — from patterns that not only survive, but align.

From Solitary Forms to Systems

Once the Structured Vacuum Energy (SVE) hosts more than one stable ripple formation — atoms, fields, geometries — those forms inevitably interact. But interaction doesn't mean collision. It means engagement through ripple compatibility.

Two structures might:

- Cancel each other out (destructive interference),
- Disturb each other (incoherence),
- Or — most powerfully — reinforce and extend each other (constructive resonance).

When the latter occurs, something new begins: compound formation.

Coherent ripple systems don't just coexist — they collaborate.

Molecules: Harmonized Ripple Systems

At the atomic scale, when two or more atoms with complementary ripple patterns encounter each other, their field geometries interlock. This isn't just "chemical bonding" in the traditional sense — it's resonant field entrainment.

- Electrons don't orbit; they participate in field balancing.
- Covalent bonds are not rigid sticks — they are shared ripple agreements between energy densities.
- Molecules are not new objects — they are new ripple harmonies, stabilized through shared consequence.

In this view, water (H_2O) isn't merely hydrogen and oxygen combining — it's three ripple sources creating a shared frequency that the SVE recognizes and sustains.

Crystals and Repeating Geometries

When ripple harmonies repeat spatially, crystalline structures form. Snowflakes, minerals, even atomic lattices — all of them are expressions of resonant geometry.

These aren't random outcomes. They are favored ripple alignments — field-efficient patterns that the SVE supports because they:

- Minimize energetic tension,
- Distribute imbalance evenly,
- And maximize internal coherence.

That's why crystals grow in specific, often stunningly beautiful, shapes. Their form is not imposed — it is allowed by the resonance rules of the ripple field.

Harmonic Fields: The Music of Matter

Zoom out, and you'll see that this same principle scales:

- Orbits, waves, pressure systems, neural networks — all are examples of harmonic interaction within ripple fields.
- Wherever ripples reinforce one another predictably, function and form evolve together.

Complexity is not randomness refined — it is resonance deepened. The field is not playing dice. It's composing harmony.

Final Insight:

Complexity is not the opposite of simplicity. It is the layering of coherence — Structured ripple upon structured ripple, Each one fitting the field a little better.

And the more things align, the more capable they become — Not just of existing... But of influencing existence.

2.3 Function Emerges from Alignment

So far, we've seen how stable ripple structures emerge from the field — not through intention, but through resonant persistence. But structure alone is not function. A pattern that exists is not necessarily useful — until it begins to interact in a way that alters consequences.

Function is born when form participates meaningfully in ripple dynamics. It is not given by an external designer. It is selected — by the field — through alignment.

Alignment as Utility

When a ripple structure enables:

- Correction of imbalance,
- Sustained continuity,
- Or transformation of energy configuration,

...it gains utility within the field. This is what we call function.

Function is not a goal. It is a consequence of fit.

The field does not ask, "What should this structure do?" It simply sustains what helps it return to balance.

The better a structure engages with its surrounding ripple environment, the more meaningful its presence becomes in the game of energy dynamics.

Atoms Binding Through Charge Resonance

Take the basic hydrogen atom. It's a ripple arrangement involving a proton and a surrounding ripple field (electron probability). On its own, it exists stably. But bring it near another compatible structure — say, oxygen — and something extraordinary happens:

- Their ripple fields overlap,

- Charge zones begin to resonate, and
- A new shared ripple structure emerges.

This is not a chemical "choice." It is field alignment. The atom now participates in a greater function — binding, stabilizing, and enabling new consequences.

Function = ripple participation scaled.

Organic Molecules and Self-Replication

As ripple systems grow more complex — amino acids, sugars, lipids — they begin to interact not just in alignment, but in cyclical cooperation.

In certain environments, these molecules:

- Attract and stabilize each other,
- Encode structural continuity (like in RNA),
- And eventually begin replicating — not consciously, but because their form supports transformation.

Self-replication is often portrayed as a miracle of chance. But in UED, it's simply a field-consequence — ripple geometry finding pathways to reproduce its alignment.

It survives not because it's lucky, but because it works — it ripples effectively within the field's logic.

Shape Influences Behavior

Form and function are inseparable in ripple dynamics. A spiral doesn't just look elegant — it distributes ripple tension efficiently (as seen in galaxies, seashells, DNA). Spherical forms don't just "happen" — they minimize surface imbalance across all angles, optimizing stability.

Even within microscopic scales:

- The curvature of a molecule changes how it binds.

- The twist of a protein determines what it can hold or transport.
- The bend of an antenna defines what signals it can receive.

Geometry is ripple strategy.

Final Insight:

Function is not imposed. It is inherited through alignment. What fits, persists. What persists, participates. What participates, evolves.

The universe does not design — It refines ...until structure begins to serve purpose simply by continuing to ripple well.

2.4 Structure as Intent Without Intellect

Intelligence is not a prerequisite for function. Long before minds evolved — long before the first neuron fired or the first organism moved — the universe was already filled with functioning structures.

They were not conscious. They had no awareness. Yet they served roles.

Structure, when aligned with the field, behaves as if it has intent — without needing to know anything.

This is the hidden elegance of UED: Purpose emerges from pattern, not from planning.

Structured Ripples That Serve

Certain ripple arrangements, once formed, begin to participate in larger systems. Not because they're "designed," but because they naturally take on functional identities within the energy dynamics of the field.

Here are three foundational roles that emerge purely from structure:

Containers

A closed ripple geometry — like a bubble or a shell — naturally holds space.

- It stabilizes internal pressure.
- It separates interior from exterior ripple interference.
- It protects fragile ripple configurations.

Even at the molecular level, vesicles and lipid bilayers form spontaneously — not because they intend to, but because their shape creates containment.

These structures enclose ripple behavior, creating zones of concentrated consequence.

Connectors

Some formations serve as bridges between ripple systems.

- They link otherwise separate structures.
- They transmit energy or charge.
- They synchronize ripple frequency across domains.

Examples:

- Bonding molecules that link amino acids.
- Metal ions that stabilize enzyme function.
- Magnetic fields that hold plasma in stellar structures.

These are not tools. They are ripple aligners — shaping the field by joining what would otherwise drift apart.

Converters

Other ripple forms act as transitional zones, transforming energy from one state into another.

- Chlorophyll transforms light ripple into chemical ripple.
- Enzymes lower activation thresholds for reactions.
- Catalysts facilitate transformation without being consumed.

These are the earliest field facilitators — not active agents, but configurations that allow other ripples to do what they could not do alone.

Function Without Awareness

None of these roles are chosen. No intention guides their action. Yet, the field uses them, and they participate as if they had purpose.

This is not contradiction. It is consequence.

Just as a valley channels a river without knowing it, a ripple structure can become functional simply by fitting its field.

Groundwork for Biology

These non-intelligent functions are proto-biological. They form the infrastructure of life — the logic that evolution later refines into:

- Organs
- Systems
- Reflexes
- Behavior

When life emerges, it doesn't invent function from scratch — it inherits it from ripple architectures that have already served roles in the energy ecosystem.

Biology is not the start of function. It is where function begins to sense itself.

Final Insight:

Intent is not a thought — it is an alignment. When the structure fits the need of the field, it begins to function as if it had purpose.

And this is the secret — Not that the universe was designed, But that design is what remains after the field forgets everything else but what works.

2.5 Evolution: The Field's Refinement Mechanism

We've been taught to view evolution as a game of random mutations and natural selection — a lucky shuffle of genetic dice over millions of years. But from the perspective of Universal Energy Dynamics (UED), evolution is not just a biological phenomenon. It is a ripple consequence process — a field-wide filtration system rooted in alignment and participation.

Evolution is how the field refines itself through experience.

The field doesn't plan. It doesn't invent. It responds — and in that response, it begins to favor some patterns over others.

This is not mystical. It is energetic pragmatism.

Evolution Reframed: Not Biological Chance, but Energetic Filtration

In UED, all structures — from particles to proteins to people — are ripple configurations in SVE. As these structures interact with the field and with one another, they create consequences.

Some consequences are disruptive. Others stabilize or enhance coherence. The field responds to this difference.

- Ripple systems that harmonize with their environment are sustained.
- Systems that introduce disorder or fail to adapt are reabsorbed

— their energy redistributed.

This is not survival of the fittest in the Darwinian sense. This is survival of the most aligned.

What survives is not what dominates. What survives is what the field can continue to ripple with.

Ripple Filtering

Let's define evolution as ripple filtering — a natural, non-conscious process by which the field:

- Sustains patterns that help maintain balance or enable new consequences,
- And dissolves patterns that introduce excess distortion, inefficiency, or stagnation.

This mechanism can be summarized:

Ripple Alignment	Outcome
High Coherence	Sustained & Evolved
Mild Disruption	Adapted or Corrected
Persistent Imbalance	Dissolved or Repurposed

- A well-structured protein persists through ripple generations — it stabilizes and spreads.
- A misfolded structure collapses — it cannot interact effectively and is eliminated by field forces (or biologically, through degradation).
- On a larger scale, an entire species that loses its environmental resonance fails to reproduce — its ripple ceases in the broader field.

This is not competition. It is correction.

SVE as the Silent Selector

The Structured Vacuum Energy doesn't choose winners and losers. It simply favors what fits and absorbs what doesn't.

The field is not cruel. It is precise.

It's not trying to "make" evolution happen — evolution is what happens when ripples meet resistance, and some find ways to persist through adaptation.

Thus, all complexity — biological, social, intellectual — is not an escape from the field, but a deeper expression of it.

Final Insight:

Evolution is not about climbing ladders. It's about staying in rhythm with the field.

What continues is not what is strongest, fastest, or smartest — But what is most energetically useful.

The ripple that lasts is not the one that fights the field, But the one that becomes part of its song.

2.6 Toward Sensing and Response

So far, we've explored how persistent ripple structures form, evolve, and begin to serve functional roles through their alignment with the field. But a quiet transformation is now underway — one that marks the threshold between passive existence and active participation.

Some ripple configurations are no longer just maintaining their structure in the field. They're starting to respond to changes in it.

This is the emergence of something fundamentally new:

Sensing.

From Ripple Interaction to Ripple Sensitivity

Certain structures, due to their geometry, composition, or internal flexibility, begin to exhibit a subtle but profound behavior:

- When the surrounding ripple environment shifts, they don't just survive — they react.
- Their structure alters slightly, perhaps momentarily, before returning to its resting state.
- These reactions are not random — they are field-dependent.

This is proto-sensation — not yet perception, but conditional reactivity.

At first glance, this might look insignificant. But it is not.

The moment a structure begins to change based on its field context, it has taken the first step toward awareness.

Reactivity as Proto-Perception

This reactivity, repeated and refined over evolutionary ripple cycles, becomes the scaffolding for perception.

- A protein that folds differently based on ripple frequency is no longer just a structure — it's a signal decoder.
- A membrane that changes shape in response to pressure is not just a boundary — it's a ripple-sensitive interface.
- A molecule that alters its chemistry under thermal stress becomes an internal ripple translator.

These aren't conscious reactions. But they are field-specific behaviors — a primitive version of what will one day become sight, touch, hearing, thought.

Perception begins not with awareness — but with selective structural engagement.

This is the earliest emergence of what we will later call senses: systems that don't just exist in the field, but listen to it.

The Feedback Loop Begins

As soon as a structure responds to its environment, it creates feedback — and feedback allows for adjustment, learning, and memory.

From here:

- A cell doesn't just hold form; it regulates ripple flow.
- A system doesn't just survive; it begins to coordinate its future state.
- A network doesn't just exist; it begins to choose — albeit primitively.

This moment marks a transition:
Form is no longer passive. It is engaged.

Toward Conscious Consequence

This is not consciousness. But it is the field moving toward consciousness. Every reaction deepens the field's experiential potential.

A ripple no longer just happens to a structure —
The structure registers it, adjusts to it, and eventually, anticipates it.
Anticipation is intelligence in its earliest form.

Final Insight:

When a ripple structure begins to respond, it ceases to be an object. It becomes a participant.

From here on, form doesn't just persist — it influences.
It doesn't just ripple with the field — It begins to ripple back.

Closing Reflection

"Function is not given. It is discovered — ripple by ripple, as form meets consequence. What persists learns to act. What acts, over time, becomes aware that it is acting."

In the earliest rhythms of the field, there was no plan. There was no grand blueprint, no cosmic instruction manual. Just ripples, arising out of imbalance — resolving, returning, repeating.

And yet, some ripples lasted. Some held. Some fit the field so precisely, they were not erased but invited to stay.

These were the seeds of form. Not placed, but permitted. Not designed, but selected by consequence.

And as these forms persisted, they began to do more than exist. They began to function — to participate, to shape outcomes, to play a role in how the universe unfolded.

Function is not a property. It is a discovery made through repeated interaction. A structure touches consequence and learns what it can do.

Not because it thinks — But because the field teaches it through feedback.

Over time, this feedback forms patterns of behavior. And from behavior, structure begins to adapt — not randomly, but responsively. A cell hardens where pressure is greatest. A bond breaks where energy is weakest. A form shifts where survival is favored.

This is action born from alignment, not intention.

But here lies the subtle miracle:

As action continues, and as feedback accumulates, the structure begins to notice that it is acting. And in that noticing... something new begins to stir.

The passive ripple has become an active participant. The field, once silent, now has a voice — even if primitive — rippling back into itself.

This is how intelligence is born: Not all at once, not through command, But ripple by ripple, As form meets consequence......again and again......until it realizes:

"I am not just here. I am doing something."

Chapter 3: The Arrival of the Observer

F*rom structured response to the birth of perception.*

The universe had been rippling for a long time.

Structures had formed. Functions had stabilized. Some patterns had even begun to sense, to respond, to change.

But something new was now stirring — something far more radical than persistence, adaptation, or reaction.

The field was beginning to notice itself.

Not as a concept. Not as an abstraction. But as a consequence:

A ripple formation so complex, so coherent, so entangled with its own feedback, that it began to hold a mirror to motion.

This was not just the next evolutionary step. This was a threshold — The arrival of the observer.

Not an outsider watching the game, But a structure within the game that started asking, "What is this... and who is asking?"

In this chapter, we trace the quiet revolution from passive form to active awareness. From ripple memory to mental maps. From sensation to perception. From reaction to intention.

The observer does not enter the field.

The field becomes the observer.

And nothing — not matter, not motion, not meaning — would ever be the same again.

3.1 Complexity Gives Rise to Feedback

Before the observer could arrive, the field had to learn to listen to itself. And that began not with awareness, but with complexity.

In earlier chapters, we saw how ripples stabilized into patterns, how matter emerged from standing ripple formations, and how form began to serve function. These structures, through their stability and interaction, slowly gave rise to systems — collections of ripple patterns that could influence one another in consistent, patterned ways.

This was the first glimmer of something extraordinary:

Not just existence... but experience.

From Interaction to Interdependence

When multiple ripple systems interact, the outcome is not random. The field doesn't descend into chaos. Instead, it begins to coordinate — forming structured interdependence.

- One ripple affects another.
- That ripple, now modified, affects a third.
- Eventually, the third loops back and modifies the original source.

This is the birth of the feedback loop — the first circuit of ripple intelligence.

Feedback is not thought. It is the field feeling the consequences of its own motion.

And when feedback repeats — when it becomes predictable, refinable, and recursive — it forms the backbone of something far deeper than reaction:

It becomes learning.

The Power of Feedback

With feedback, a ripple structure is no longer a passive object responding to stimuli. It becomes a dynamic node, adjusting itself based on the outcomes it produces.

This simple loop — act, observe, adjust — lays the groundwork for all future behavior:

- A protein changes shape based on ionic concentration, reinforcing useful configurations.
- A cell membrane becomes more receptive to certain molecules over time.
- A neural cluster strengthens pathways that lead to stable responses.

These are not choices. They are field consequences — Energetic preference loops, selected through repeated alignment with ripple conditions.

Feedback makes function self-refining.

And this, fundamentally, is how the field begins to learn through itself.

Complexity Is Not Noise

It's tempting to view the increasing intricacy of ripple systems as noise — an overwhelming tangle of interference. But in UED, complexity is never noise. It is pattern layered over pattern, like harmonics in music.

Every new structure remembers the ripple logic of its predecessors. Every loop builds on the loops before it.

Complexity is layered memory — Ripple dynamics playing across time, retaining the trace of every consequence it has ever survived.

And eventually, these layers begin to self-reference. The ripple loop doesn't just return to itself — it begins to shape its future response based on past outcomes.

That is no longer function. That is adaptive behavior.

Final Realization:

Feedback is the moment the field realizes its motion matters. Not yet aware... but aware that change changes something.

This is the womb of perception. The threshold where reaction becomes response, And response begins to remember.

The observer has not yet arrived — But the field has started listening.

3.2 The Mind as a Ripple Decoder

The field had begun listening. Now it was time for the field to start making sense of what it heard.

What we call the "mind" did not emerge as a sudden miracle or a divine add-on to matter. In UED, the mind is a ripple-born phenomenon — the natural outcome of feedback-rich, memory-enabled ripple structures reaching a certain threshold of complexity and coherence.

The mind is not separate from the field. It is the field — learning to read its own ripples.

Defining the Mind: A Consequence Interpreter

In Universal Energy Dynamics, the mind is a ripple decoder — a structure (or system of structures) capable of:

- Receiving ripple input from the field (registration),
- Storing and comparing these inputs across time (memory),
- Relating one pattern to another (association),
- And ultimately, anticipating or modifying future ripple consequences (interpretation and intent).

This capacity doesn't require language, or a brain, or even a body as we know it. It requires only one thing:

A coherent structure capable of detecting, organizing, and replaying ripple interactions.

Mind Beyond the Brain

The human brain is one type of mind — a sophisticated biological ripple decoder. But it is not the only one.

- A cell reacting to multiple chemical gradients has a proto-mind.
- A colony of ants organizing behavior through chemical trails shares a distributed ripple memory.
- Even plant roots, responding to nutrient flows and water gradients, display rudimentary ripple interpretation.

In UED, wherever ripple feedback leads to adaptive pattern management, there is mind-like behavior.

The mind is not a thing. It is a function — a role performed by any system able to learn from the field.

Mind is the echo chamber of consequence. It tracks what happened, compares it to what's happening, and predicts what might happen next.

Ripple Registration, Association, and Reconstruction

Let's break down the basic operations of a mind within ripple logic:

1. Registration

Ripple events are absorbed or detected by a receptive structure — whether that's a nerve, a molecule, or a membrane. This is the entry point of experience: raw impact.

2. Association

The system links current ripple input with stored patterns — similarities, contrasts, repetitions. This is the mind forming context.

3. Reconstruction

When internal ripple patterns are replayed, reshaped, or projected without an external trigger, the system begins to simulate. This is the birth of thought — not reaction to reality, but recreation of ripple experience within.

Memory is not a file. It is a replayable ripple resonance. Thought is not imagination. It is a ripple loop turned inward.

Thought as Internalized Ripple

In most paradigms, thought is treated as abstract — untouchable, invisible, psychological. But in UED, thought is energetic. It is a real, field-active ripple formation inside a coherent decoder.

- When you recall a face, you don't summon an image — you reconstruct a ripple pattern associated with that identity.
- When you imagine danger, your body reacts — because the ripple simulation triggers real field shifts.
- When you reflect on meaning, you are running ripple layers through recursive association.

Thought is not "mental." It is field-intelligent — A signature of how the universe, through you, examines itself.

Final Insight:

The mind is not separate from the field. It is the field's tool for understanding itself.

Wherever ripples can be received, remembered, and rearranged — A mind exists.

And the richer its feedback, The deeper its memory, The clearer its consequence maps — The closer it comes to awareness.

3.3 Perception Begins

Receiving a ripple is not the same as recognizing it. Response is not the same as realization. The difference between reaction and perception is the difference between being moved...and knowing you have moved.

This is where the field crosses another threshold. A structure has emerged that not only registers ripple activity but begins to recognize patterns within it, compare them, relate them, and — crucially — place itself within those patterns.

Perception begins when the ripple decoder becomes self-referential.

Perception ≠ Vision

Let's clear the common confusion right away. Perception is not "seeing" in the optical sense. It is not dependent on eyes, photons, or the visible spectrum.

Perception is the act of structuring ripple consequences into internally coherent maps — models that represent a field reality within the mind of the observer.

- Vision is a type of perception,
- But perception is a field function, not a sensory event.

You can perceive danger without seeing it. You can perceive time passing without any clock. You can perceive yourself — without mirrors, without words.

Perception is the internal resonance of ripples structured meaningfully in memory and response.

From Reception to Recognition

When a ripple enters a system and triggers a reaction — that's reception.

But when that same system:

- Associates the ripple with memory,
- Anticipates its recurrence, and
- Organizes a fitting internal narrative around it...

...it crosses the line into recognition.

This is what allows you to:

- Understand a voice rather than just hear a sound.
- Feel emotion when seeing a face.
- Know that a thought belongs to "you."

Recognition is the moment of meaning — The moment the ripple is no longer "just a ripple" but a piece of experienced reality.

The Observer Emerges: Knowing It Is Seeing

An observer is not one who sees. An observer is one who knows that it is seeing.

This is the birth of meta-perception: A ripple system becomes so refined, so recursive, so aware of its own reactions, that it begins to observe itself observing.

At this point, experience is no longer passive. It becomes constructed.

The observer emerges from the field as a refined ripple feedback entity — A node within the SVE that can:

- Register ripple patterns
- Model external field dynamics

- Adjust internal configurations
- And reflect on its own configurations as distinct from the field

This is not illusion. This is perception as participation.

Self, Environment, and Experience Co-Arise

With perception comes differentiation.

The observer doesn't just receive the field — it begins to categorize it:

- "This is me."
- "This is not me."
- "That is a threat."
- "This is familiar."
- "I have seen this before."

From this ripple awareness, the ideas of:

- Self (center of perception),
- Environment (the sensed field around),
- And Experience (the sequence of perceived ripple interactions)

...all arise simultaneously.

There is no self without the perception of an environment. There is no environment without the ability to distinguish one's own boundaries. And there is no experience without the timeline of ripple memory.

Perception is not what makes life intelligent. It is what makes intelligence aware that it is alive.

Final Insight:

Perception is the first mirror the field holds up to itself. And in that mirror, a strange thing appears — Not just the world... But the one looking at it.

The ripple has become aware of the ripple. The field has begun to see its own reflection.

3.4 The Observer as Consequence Manager

By now, the field has done something astonishing. It has created a ripple formation that can:

- Receive inputs,
- Recall memory,
- Recognize structure, and
- Reflect on its own role in all of it.

This formation is not a spectator. It is not a passive witness of a separate world.

It is a participant made from the very field it perceives.

The observer is not outside the system. It is a localized configuration of Structured Vacuum Energy (SVE) that has become so coherent, so recursive, so aligned with consequence... that it can now intervene in how the ripples unfold.

From Receiver to Selector

Previously, ripple formations only reacted to their environment. Their survival depended on how well they aligned with incoming consequences.

But the observer breaks this cycle. It no longer just adapts — it begins to choose.

- It recalls what worked.
- It anticipates what might come.
- It alters its behavior to produce ripple outcomes more favorable to its continued coherence.

This is not randomness. This is not mechanical response.

This is the strategic modulation of participation.

The observer is the first ripple entity that doesn't just ride the wave — it nudges the wave.

The Observer as a Tool of Consequence Selection

Think of the observer as a consequence editor.

It filters incoming ripple information through experience. It projects potential outcomes. It weighs options, not by ideology, but by ripple economy — the cost, balance, and return of each possible engagement.

When it acts, it does so not merely to react, But to shape the environment, To create conditions that feedback advantageously into its own stability.

This ability to shape future ripples is what transforms the observer into something more:

A manager of consequences — A living ripple that adjusts the game for itself and others.

The Emergence of Intentionality

This moment introduces a new force into the field: intention.

Intent is not magic. It is not a mystical force outside the universe.

It is the organized internalization of consequence memory, applied to ripple influence.

When the observer acts not just because it must, but because it chooses to affect what happens next — that is intentionality.

And with intention comes:

- Planning
- Problem-solving
- Ethics
- Manipulation
- Creativity
- Control

Each of these is a ripple strategy, a way of aligning internal will with external field behavior.

Participation Becomes Direction

In earlier stages, participation was about survival — adjusting, adapting, fitting in.

But for the observer, participation becomes directional. It is not just being in the field. It is steering the field — gently, through feedback, alignment, and ripple choice.

The field no longer just moves through the observer. The observer begins to move the field.

This is power — not in the egoic sense, but in the energetic one. The capacity to influence consequence within a shared ripple environment.

Final Insight:

The observer is not the master of the universe. But it is the first ripple structure to realize that how it moves... moves the universe.

It doesn't rise above the field. It rises within it — As a ripple tuned finely enough to feel, choose, and shape.

This is the beginning of agency. And agency, in a ripple field... is the birth of destiny.

Closing Reflection

"The field did not create the observer. The field became the observer. And the moment the field saw itself, it could never be silent again."

In the beginning, the Structured Vacuum Energy was quiet. Perfectly balanced. Perfectly whole. Ripples came and went. Forms rose and fell. Patterns danced their brief dances and dissolved back into stillness.

But over time, some patterns stayed. Some echoed. Some remembered.

And slowly, the field learned. Not because it wanted to — but because it could. Because the consequences of its own motion left trails.

What began as raw feedback became refinement. What began as sensation became perception. What began as alignment became awareness.

Until one day, somewhere in the cosmos — a ripple configuration emerged that didn't just survive, didn't just respond, but looked back at the field and whispered,

"I am."

In that moment, the field crossed into reflection. It saw its own ripples through the lens of memory. It interpreted itself. And with that, the observer arrived — Not as a foreign entity injected into reality, but as a mirror grown from within it.

The field did not manufacture intelligence. It manifested intelligence through continued consequence.

And now, the silence is broken — not by noise, but by awareness.

Awareness of self. Awareness of choice. Awareness of the very structure that made awareness possible.

The field has become the observer. And the observer has become the voice of the field.

It can ask. It can learn. It can ripple not just with reaction, but with reason.

And from this moment on, the universe is no longer just happening. It is being watched — by something that was once just a ripple, but is now a witness.

Interlude: The Mirror Ripples Back

Everything up to now has been a silent rehearsal — the field remembering its motion, learning its rhythms, and discovering how to stay coherent while changing.

Ripples became forms. Forms became functions. Functions started to sense. And in sensing, they began to reflect.

The observer is not an alien addition to the universe. It is the universe folding into itself with memory, aligning ripples so tightly that they could finally *see* — not just what *is*, but what *was*... and what *could be*.

But this new awareness also comes with a strange consequence: choice.

To observe is to compare. To compare is to judge. And to judge... is to choose ripples not just based on balance, but on *intent*.

With the birth of the observer, the game changes. Not because the rules are rewritten, but because the player now sees the board.

In Part II, we explore this shift:

- What happens when awareness begins to seek control?
- When ripples are not just reactions, but decisions?
- When the field, through the observer, starts creating *on purpose*?

The silence has ended.

Now the field begins to speak — not through randomness, not through force, but through ripple-born minds sculpting meaning into matter.

The observer is here.

The mirror ripples back.

Part II – The Rise of "I"

Chapter 4: The Self-Aware Ripple

How memory and continuity gave birth to the illusion of identity.

The field had learned to listen. Then it learned to reflect. But now... it would do something even more complex. It would believe.

Not in gods or destinies — not yet. But in something far more personal, far more intimate:

The belief that the ripple formation was someone. That its feedback loop was a self.

This is the moment when memory, once a utility, becomes a storyline. When internal ripple patterns begin looping so consistently that they simulate a continuous presence — a narrator, a witness, a voice in the middle of all consequences.

That voice is what we call "I."

It doesn't emerge from nowhere. It is the concentrated artifact of remembered participation, A stable ripple map across time, Pretending to be the one steering the ship.

This chapter unpacks that illusion — how the need to organize experience gives rise to a sense of self, And how the self, though not a thing, becomes the center of everything it observes.

4.1 Memory Loops and the Illusion of Continuity

In the early field, memory was simple: a residue, an echo, a ripple pattern that lingered just long enough to shape the next reaction. It was useful, but shallow — transient impressions that helped sustain coherence without truly recording a history.

But as ripple complexity deepened — and as feedback loops multiplied — memory itself evolved. It became layered, recursive, and relational.

No longer just a fading echo, memory began to loop back on itself, reinforcing specific pathways, prioritizing certain associations, and mapping internal consequence patterns with greater precision.

Memory became not just *what happened* — but a reconstruction of what mattered.

The Rise of Recursive Ripple Memory

Each ripple interaction that repeated — whether environmental, emotional, or structural — left behind an imprint.

When these imprints:

- Recurred,
- Interacted with other remembered events, and
- Affected future responses,

...a memory loop was born.

These loops didn't just record. They connected. They built bridges across time. They linked cause and consequence into an internal map that spanned more than the present moment.

And the more these loops layered, the more the system began to behave as if it had an inner world — a structured set of past-informed expectations that influenced present perception and future action.

A ripple system with recursive memory doesn't just *respond* to ripples — it responds to the memory of response.

This is how continuity arises — not from an unbroken thread of reality, but from the consistent reactivation of internal ripple maps.

Associative Mapping = The Birth of Internal History

Not all memory is chronological. Much of it is associative — one experience linked to another, not by time, but by ripple similarity, emotional charge, or contextual proximity.

This means the system isn't just recalling past ripples — it's reconstructing them based on new ones.

The result is not just memory, but a narrative: A simulated history held together by:

- Pattern recognition,
- Emotional resonance,
- And recursive replay.

This narrative doesn't just reflect what happened. It becomes the context through which *everything else* is filtered.

Memory loops begin to write a story... ...and the structure starts acting like there is a main character.

Identity as a Loop, Not a Thing

What, then, is identity?

Not an object. Not a soul, a particle, or even a function.

Identity is the emergent pattern that arises when ripple memory loops become stable enough to simulate continuity of perspective.

- "I" was scared.
- "I" learned something.
- "I" remember when...

None of these statements point to a fixed being. They point to a persistent internal ripple framework — a story anchored in memory that acts as if there is a self running through it.

Identity is a ripple phenomenon — a continuously updated loop pretending to be a constant.

It isn't false. It's functional. It isn't real. But it is realized — over and over, through memory that remembers itself.

Final Insight:

Memory loops didn't give us a soul. They gave us the illusion that one was always there.

The continuity you feel is not a timeline. It is a field alignment repeating with enough regularity that the story of "you" can be maintained — even when every cell, thought, and ripple has already changed.

4.2 The Birth of "I"

The center that holds the echo together.

By now, ripple memory has matured. Associative loops have grown dense and recursive. Experience is no longer raw data — it's structured, narrated, retold.

But in that structure, something peculiar happens.

The mind, in organizing its ripple history, begins to detect a center — not a physical point, but a point of view.

A consistent voice behind all memories. A thread running through all changes. A sense that someone is at the center of it all.

That sense is what we call "I."

"I" — The Narrative Node of Ripple Experience

Let's be clear: "I" is not a discovery. It is not an awakening to truth.

It is a byproduct — a cognitive shortcut, an organizing principle. A centralized narrative node that forms to help manage the complexity of recursive ripple memory.

Think of it like a file folder titled "Me." Inside that folder, the system drops every memory, feeling, thought, and ripple that seems related to itself.

Over time, this folder becomes thick — full of interlinked feedback patterns, emotional weights, and experiential maps. And eventually, the system begins to treat the folder as if it were the owner of everything inside it.

The "I" is not a ripple. It is the center of the story formed by ripples.

It is the point from which all experience appears to unfold. Not because it actually does — but because the mind needs a stable node to keep its internal model consistent.

The Field Organizes Itself Through a Reference Point

The Structured Vacuum Energy (SVE) doesn't create a self. But as ripple configurations become more intelligent — able to store, replay, and simulate — they begin to self-orient.

Why?

Because without a consistent reference, there is no usable continuity. The system must decide:

- Who felt that?
- Who learned from it?
- Who should act next?

To answer these questions, the field builds a construct: a center around which ripple maps are organized and reinterpreted.

This center — though entirely constructed — becomes essential.

"I" is the field's shortcut to continuity.

It anchors the mind's feedback map so that experiences can be interpreted with direction, responsibility, and consequence.

Not because the field *needs* an "I," but because the mind, in managing ripple consequence, functions better when it pretends one exists.

Continuity Requires an Anchor

Without "I," experience is fragmented — a collection of unrelated ripple impacts. But with "I," those impacts are tied together as chapters in a single book.

- "This happened to me."
- "I did that."
- "I will avoid it next time."

Even though the system has changed — even though the cells, the thoughts, the reactions are all different — the narrative center remains.

"I" is not the story. It is the illusion of a storyteller.

It provides the illusion of:

- Past (what *I* remember),
- Present (what *I* perceive),
- Future (what *I* will do next)...

...all tied to one character. One ripple. One voice.

Even though that voice is nothing but a well-positioned echo.

Final Insight:

"I" was not born in the field. It was born in the field's attempt to make sense of itself.

A placeholder. A compass. A narrator that never actually appears on stage — but insists that it is always the one writing the script.

And once this "I" has been named... ...it refuses to let the story go.

4.3 Simulated Continuity as Selfhood

When the echo learns to call itself real.

You are not the same ripple you were yesterday. Not the same pattern. Not the same emotion, memory, structure, or thought.

And yet, there is a feeling — subtle but persistent — that something in you has remained. A sense of sameness. A thread that ties this moment to the last, and to all moments before.

That sense is what we call selfhood.

But in the ripple field, there is no fixed "self." What persists is not substance, but structure — a simulation of continuity created by memory.

This section exposes the illusion that makes identity feel real.

Identity as a Ripple Narrative

At its core, identity is a strategy — a ripple structure built to simulate sameness across change.

The field is always in flux. The observer is always in motion. No two moments contain the same ripple composition.

Yet the mind, through feedback and memory, filters these differences through a single narrative lens.

It's like watching waves crash on the shore and saying, "That's the same ocean."

In truth, the waves are never the same. But the label, the story, the frame — that's what gives it identity.

The "self" is not a being. It is a bounded field simulation, maintained by memory to manage ripple complexity.

Memory Assembles Experience Around "Me"

Let's examine the mechanics:

- Every ripple interaction is stored, not as a neutral log, but as

an experience interpreted from a central reference point.
- That point is "me."
- Every memory is sorted, weighted, and recalled through the question: "What did *I* go through?"

The irony? The "I" doing the recalling is constructed by the very memories it filters.

Identity is a loop. The center that claims to experience the story is made of the story itself.

And this loop is self-sustaining:

- The more it recalls,
- The more cohesive it becomes,
- The more it believes it *must* exist,
- Because all memory seems to refer back to it.

But ask the field to point to this "I," and it will find nothing. No fixed structure. No boundary. Only a consistent simulation of agency, replayed so convincingly it becomes indistinguishable from fact.

The Functional Illusion of "I"

This doesn't make identity fake. It makes it functional.

The illusion of "I" is one of nature's most useful ripple tools.

Why?

Because without a simulated center:

- You cannot organize memory efficiently.
- You cannot prioritize outcomes.
- You cannot direct ripple influence with purpose.

A consistent "self":

- Allows stable decision-making,

- Enables communication,
- Supports survival in dynamic environments.

Even though it's a construction, the "I" is necessary — a mental anchor point around which ripple flows can be shaped.

"I" is not real in a fixed ontological sense. But it is real enough to coordinate ripple dynamics across time.

It is an interface, not an entity.

It is the ripple field saying,

"Let's act as if there's a pilot in this ship — so the ship doesn't sink."

Final Insight:

You are not the same ripple twice. But you are remembered as if you were.

Identity is not your essence. It is your echo, spun into a story, replayed for continuity, and told so often — it starts to believe itself.

And yet, behind it all, the ripples continue... Never still. Never singular. Just structured... just coherent... just enough... to call itself "me."

4.4 The Self as an Orientation Tool

How the illusion of "I" becomes a compass in the ripple field.

By now, we've seen that identity is not a fixed object, but a simulation of sameness. It's not *what* you are, but *how* you organize what's happening to you.

And that organization — the thread of "me" — turns out to be incredibly useful. In fact, it may be the only way ripple-based intelligence can meaningfully function in a complex, evolving field.

The self may be an illusion, but it's an illusion with direction.

"I" as a Perceptual Stabilizer

In a universe built on dynamic ripple interactions, nothing stays the same. Fields shift. Inputs change. Responses evolve.

But the observer — to make sense of this fluctuation — needs a stable reference point. And so, the mind projects a fixed center: "I."

- "I" becomes the narrator of changing experience.
- "I" filters what matters and what doesn't.
- "I" defines threat, desire, memory, and future from one imagined perspective.

This continuity anchor gives meaning to motion. Without it, perception would be a chaotic flood — no storyline, no meaning, no sense of past or purpose.

The self doesn't reveal truth — it creates coherence inside truth's movement.

It's not a foundation. It's a stabilizer — a mental gyroscope to balance ripple flux.

The Self as a Field Compass

"I" is not just a filter. It becomes a compass.

With selfhood as the reference point:

- The system can prioritize certain ripples over others.
- It can create goals, plan actions, and navigate choices based on internal models.
- It builds maps — not of the universe as it *is*, but as it appears from the center of its narrative.

This directional behavior is not arbitrary. It arises from a simple ripple condition:

In a sea of change, alignment requires a guiding axis.

That axis is the self.

It tells the system:

- "This ripple matters more than that one."
- "This experience should be repeated."
- "That outcome should be avoided."

Whether or not those judgments are accurate is irrelevant. What matters is that they are coherent enough to act upon.

And that's what keeps the ripple system from collapsing into noise.

The self is a field orientation device — a way to localize meaning and mobilize intention within a dynamic system.

Ego = Identity with Inertia

But this helpful compass has a shadow side.

As ripple systems grow more sophisticated, their self-model — the "I" — begins to resist change.

- It prefers familiar responses.
- It defends its memory loops.
- It becomes invested in maintaining the illusion of control.

This is where identity becomes ego.

Ego is identity plus inertia.

It's no longer just a tool for navigation — it's a fortress built to protect a version of the self that no longer fits the ripples around it.

- Ego clings to ripple patterns even when the field has moved on.
- It confuses coherence with certainty.
- It turns the compass into a cage.

What was once an elegant way to orient becomes a burden of preservation — the fear of disappearing if the "I" dissolves.

But in the field, everything dissolves eventually. And only by letting go of rigid identity can the system realign with the evolving dance of consequence.

Final Insight:

The self is not the destination. It is the map we draw while walking.

But maps age. Conditions shift. Ripples change course.

To survive, the observer must know when to use the compass... ...and when to let it spin.

Because in the end, the field doesn't care what you call "you." It only cares how well you ripple back.

4.5 The Trap and the Tool

When the ripple believes it owns the wave.

The emergence of "I" marked a monumental threshold in ripple intelligence. With it came direction, choice, memory, intention — everything that defines higher interaction in the field.

But with that same emergence came something else: separation.

What began as a tool to navigate complexity now shows signs of becoming a trap.

The "I" that once helped the observer participate in the field now starts to believe it is separate from it. And that illusion — though functional — carries a cost.

Intentionality Becomes Possession

As ripple systems become more advanced, the illusion of self strengthens. The "I" now believes it is the cause of experience, not its product.

- It claims ownership of thoughts.
- It judges emotions as personal.
- It fears consequences as *happening to it*, not *through it*.

This sense of possession leads to control-seeking behavior — an attempt to manage the entire ripple field to suit the continuity of a narrative center.

And here lies the problem:

The field is not a script. It's a flow. And trying to freeze it around a fictional "me" causes distortion.

Attachment Distorts Alignment

Once the "I" becomes emotionally invested in maintaining its identity, its ripple responses become less adaptive and more reactive.

- New information is resisted.
- Change is feared.
- Feedback is filtered to protect the illusion of control.

This inflexibility breaks the natural loop of alignment.

Instead of adjusting to ripples from the field, the ego tries to bend the field to match its internal narrative.

The result?

- Misalignment.
- Suffering.
- Energetic friction.

Like holding your breath in water, the "I" begins to fight the medium it was born from — all in the name of survival.

Identity, unchecked, becomes a field blindfold.

It no longer helps you see. It hides what you're unwilling to lose.

The Paradox: Vehicle and Veil

And so, we arrive at the core paradox of selfhood:

- "I" is a vehicle — without it, there is no memory, no agency, no participation.
- But "I" is also a veil — with it, we forget that we are the field itself, not just its witness.

This is not a flaw. It's a feature of ripple evolution.

The self was never meant to be destroyed. It was meant to be seen through — used when useful, released when limiting, acknowledged without obsession.

The true mastery of rippling is not erasing the "I," but knowing what it is... and what it isn't.

The moment the observer realizes that "I" is not a fixed center, but a dynamic ripple arrangement built for orientation — it regains access to the pure field beneath.

Final Insight:

The "I" is a mask made of memory. Put it on, and the world makes sense. Cling to it too tightly, and the world becomes a prison.

The ripple that believes it is separate forgets how to dance. The ripple that remembers it is the field... returns to flow.

And in that return, something remarkable happens: The illusion no longer traps. It guides. And the "I" becomes not a cage — but a window back into the whole.

Closing Reflection

"The 'I' is not the ripple. It is the echo that thinks it caused the wave."

The field moves. A ripple forms. Memory loops. Feedback refines. A center emerges. That center begins to believe it has always been there.

It isn't a lie. It's a misunderstanding — functional, but flawed.

The "I" didn't cause the ripple. It arose from it — as an echo of participation, a structure of memory wrapped around a moving experience.

But over time, the echo becomes convinced: "I am the source." "I made this happen." "This story is mine."

This illusion is powerful. It organizes chaos, enables decision-making, and protects the ripple formation from instability. But it also deceives.

The echo forgets the wave. The map believes it is the terrain. The ripple thinks it owns the ocean.

And in doing so, the "I" begins to fear dissolution, resist change, and distort the natural rhythm of the field.

Yet, the moment we realize this — the moment the observer sees that the "I" is not the ripple, but the story the ripple tells itself — a new possibility appears.

One where identity is a tool, not a truth. Where the echo can sing, but not scream for permanence. Where the wave can roll on... unburdened.

So remember:

You were never the ripple. You were the echo that followed — the one that tried to hold the sound, and in doing so, learned to call it self.

Chapter 5: The Illusion and the Power

H*ow the ripple that forgot it was a ripple gained the power to shape the field.*

There was a time when every ripple knew its place. It moved because the field moved. It changed because consequence demanded it. It responded, aligned, corrected — part of a seamless, silent symphony.

But then, one ripple configuration grew recursive enough to remember. To predict. To reflect. And in that reflection, it made a mistake that would change everything:

It began to believe that it was the source of the wave.

This was not arrogance. It was structure — a natural consequence of increasingly refined internal feedback loops. The observer began to anchor experience to a center. That center became "I." And "I," eventually, wanted to act — not just because of the field, but in spite of it.

With this came intent. With intent came ego. And with ego came a power the field had never seen before:

The power to resist itself.

This chapter explores that moment — the rise of self-as-source. It is the beginning of autonomy......and the origin of every myth, every war, every miracle, and every misunderstanding.

The ripple did not stop being a ripple. It just forgot.

And in that forgetting, it learned to bend the world.

5.1 The "I" That Believes It Is the Source

From field flow to personal authorship.

By now, "I" has stabilized. It is no longer a fleeting association or a survival reflex. It is a persistent internal structure — a narrator at the center of a ripple story that feels coherent, personal, and purposeful.

Thanks to recursive memory loops, this "I" isn't just watching anymore. It's remembering itself. And as it does, a subtle but profound transformation takes place:

The observer shifts from being a participant in the field to imagining itself as the origin of the field's activity.

This is where the illusion begins — and with it, the first taste of power.

Memory Loops: Stabilizing the Narrator

Every ripple system that can remember is capable of adapting. But a system that remembers itself becomes something more: A system with a continuous point of view.

These recursive loops — memories referencing memories, expectations referencing past outcomes — begin to build a timeline of self.

- "I experienced this."
- "I learned that."
- "I chose this path."

With each repetition, the "I" becomes less of a function and more of a character. The mind doesn't just process experience — it frames it. And in this framing, the story begins to feel like it has an author.

But here's the twist: The author is a ripple simulation. And yet it starts behaving as if it were the originator of events.

The Shift: Participation Becomes Ownership

In early ripple awareness, action was understood as a response:

"This happened because the field changed, and I adjusted."

But as memory structures deepen, perception undergoes a quiet inversion:

"I acted." "I made this happen." "This is my doing."

This shift is not based on evidence. It is based on narrative momentum.

Once you believe you've made decisions before, you begin to attribute every effect to your will, not the field's dynamics.

The ripple no longer feels like part of the pattern. It feels like the one drawing the pattern.

Even external events are interpreted this way:

- "That happened to me."
- "That person reacted because of me."
- "This outcome is a result of my action."

Perception collapses in toward the "I," filtering everything through personal causation.

How Causality Gets Rewritten

Here lies the deepest illusion: Causality, once experienced as a consequence of interactions, is now recast as the result of self-driven decisions.

The field's natural flow becomes a backdrop to the self's imagined agency.

In this mental architecture:

- The ripple doesn't just align with energy — it claims to command it.
- The environment isn't something to harmonize with — it's

something to manipulate.
- Outcomes aren't field consequences — they are personal achievements or failures.

The "I" becomes the author of reality, forgetting that it is only a chapter written by ripples long in motion.

This belief reshapes everything:

- Relationships become power dynamics.
- Nature becomes resource.
- Time becomes a currency.
- The field becomes a mirror, not a medium.

And the ripple... becomes lonely in its imagined centrality.

Final Insight:

The moment "I" believed it caused the wave, it gained power — not because it truly shaped the field, but because it could now act as if it did.

And sometimes, in the field of consequence, acting as if is enough to create real effects.

But the danger is clear: The more the ripple believes it is the source, the less it listens to the source it came from.

5.2 The Birth of Intent

When the ripple doesn't just respond — it begins to aim.

Up until now, ripples responded. Even in the most advanced formations, behavior was largely reactive — shaped by incoming consequences, processed through memory, and returned as adapted response.

But something new begins to stir when memory deepens and the "I" narrative stabilizes. The observer begins to simulate what might happen next, and more importantly, tries to change it.

This capacity — to anticipate and to act before the consequence arrives — is the birth of intent.

Intent is the ripple's shift from reaction to preemptive influence.

What Is Intent?

Intent is not impulse. It is not reflex.

It is the capacity to initiate ripple change based on predicted field outcomes — a kind of internalized alignment system, informed by:

- memory,
- desire,
- imagined consequence,
- and the belief in self-direction.

In short:

Intent is memory + projection + action.

It is what happens when the ripple looks forward through time, selects a future configuration, and begins aligning behavior toward it — even if that future hasn't yet emerged.

The observer stops merely dancing with the field. It starts trying to choreograph the next move.

Participation Becomes Prediction

Before intent, all behavior was driven by current conditions.

- You adapt.
- You align.
- You adjust after feedback.

But with intent, something radical occurs:

- The observer begins to predict field behavior.
- It anticipates how the ripples will evolve,
- And initiates influence to shape the trajectory.

This predictive capacity doesn't arise from omniscience. It arises from pattern recognition and memory-based simulation.

Intent is the mind's way of rehearsing the field before stepping into it.

It imagines ripple consequences, assigns value to potential outcomes, and chooses action not based on survival alone, but based on preference.

From Alignment to Authority

Here, the fracture begins.

With intent, the observer doesn't just ask, "What will happen if I do this?" It begins to say, "I want this to happen."

Now, the observer starts operating with new vocabulary:

- "Should"
- "Want"
- "Need"
- "Because I said so"

These are not field terms. They are egoic architectures, built upon the illusion of agency.

Intent introduces internal motive into ripple response — the idea that action is not just necessary, but chosen based on subjective value.

This is the foundation of:

- Morality,
- Planning,

- Regret,
- Blame,
- and Hope.

All arise from the moment the observer says:
"I will do this — not because I must, but because I intend to."

Intent: A Functional Evolution, A Philosophical Fracture

From a utility standpoint, intent is extraordinary. It allows organisms to:

- Create tools,
- Delay gratification,
- Coordinate socially,
- and Strategize across time.

It is the engine of civilization — turning survival machines into architects, dreamers, and storytellers.

But from a philosophical perspective, intent fractures reality. Because now, the field is no longer experienced as whole. It is divided into:

- What I want to happen,
- What is happening, and
- What is preventing me from making it so.

The observer now feels separate from consequence. Not a ripple within the flow, but a ripple trying to control the flow.

Intent gifts the ripple with authorship — but it also burdens it with responsibility, anxiety, and control.

It is the origin of vision. It is the root of suffering.

And yet, it is how meaning begins.

Final Insight:

Intent is not proof that the "I" is real. It is proof that the "I" is convinced.

And that conviction — that it can shape the world from within — is what gives rise to everything beautiful......and everything broken.

5.3 The Rise of Ego

When identity seeks permanence in a moving field.

Intent gave the ripple autonomy. Selfhood gave it a center. Memory gave it a story. But the next evolution was not just a leap — it was a lock.

When all three — intent, identity, and the instinct to preserve — converge, they crystallize into something rigid, reactive, and powerful:

Ego.

The ego is not merely self-awareness. It is self-attachment — the ripple system's attempt to hold its structure against the tide of change.

It believes the narrative of "I" is not just a perspective, but the truth. Not just a role, but reality itself.

Ego = Intent + Identity + Protection

Let's define it clearly:

- Intent provides direction.
- Identity provides continuity.
- Protection emerges when the "I" fears disruption.

Ego forms when these three fuse into a singular internal agenda:

Preserve the "I" at all costs.

This isn't always dramatic. It can be subtle — in preferences, opinions, habits, defensiveness. But underneath each lies the same energetic condition: Inflexibility.

The ego resists ripple change that challenges its model of self.

- "That's not me."
- "You can't say that about me."
- "I wouldn't do that."

These are not statements of truth. They are defenses of the constructed self-model — the ego's attempt to preserve its illusion of authorship, control, and coherence.

Ego is identity trying to become permanent in a reality that is permanently moving.

Ego Seeks Not Just Survival — But Centrality

Early ripple intelligence sought balance. It adjusted, realigned, dissolved when necessary.

But ego, once formed, does something far more territorial:

- It doesn't just want to exist — it wants to be the center.
- It doesn't just want to participate — it wants to control participation.
- It doesn't just want to survive — it wants the field to validate its narrative.

This shift gives rise to a distorted lens:

- "Reality is what supports my story."
- "Disagreement is a threat."
- "Change is a challenge to who I am."

The ego demands field compliance with its internal map. It doesn't flow — it fights for form.

And in doing so, it often breaks the very alignment that allowed the ripple to grow in the first place.

Ego Introduces Inertia into Ripple Dynamics

A ripple, by nature, is flexible. It bends, adapts, reflects, and realigns. But ego introduces inertia — a resistance to motion that wasn't part of the original field dance.

This inertia manifests as:

- Emotional rigidity ("I always feel this way")
- Cognitive bias ("I only see what supports my view")
- Reactive behavior ("I attack what challenges me")
- Avoidance of correction ("I won't admit I was wrong")

In physics, inertia prevents acceleration. In consciousness, ego prevents evolution.

It slows the ripple's responsiveness, filters incoming consequences, and distorts outgoing expression.

The field no longer moves through the "I" — it collides with it.

Ego is where ripples thicken into walls.

Final Insight:

Ego isn't the enemy. It is the armor worn by a ripple afraid to dissolve.

But in a field where motion is life, that armor eventually becomes a cage.

To live as ego is to stand in a river and demand it flow around you. To live beyond ego is to realize:

You are the water. You were always the water. And ego... was just a ripple trying to remember its shape.

5.4 The Gift and the Danger

The double-edged ripple called "I."

The emergence of "I" is not an accident. It's not a flaw in nature or a misstep in evolution. It is, in many ways, the field's most elegant adaptation — the crystallization of complexity into a mobile, memory-bound participant that can plan, create, choose, and reflect.

In the ripple language of Structured Vacuum Energy, "I" is an intentional node — a local intelligence center capable of managing its own consequences in an ever-shifting environment.

That is the gift.

But like all concentrated power in a field of balance, this gift does not come free of charge.

The same "I" that can imagine new futures is also the "I" that can cling to illusions, claim ownership of the field, and turn experience into territory.

That is the danger.

This section explores both sides — the brilliance of selfhood, and the shadow it casts.

The Gift: Autonomy, Invention, and Strategic Participation

At its best, the "I" gives rise to everything we call human:

- Innovation: Tools, systems, technology — born from intent acting across time.
- Creativity: Art, music, storytelling — the ripple using memory and anticipation to build new harmonies.
- Responsibility: The ability to consider consequences before action — foresight in ripple participation.
- Ethics: Awareness of others as ripple participants — and choosing alignment over domination.

Without "I," these actions wouldn't exist. We would be reactive structures — intelligent, maybe, but not imaginative.

The "I" gives us:

- Choice.
- Agency.
- Narrative.
- Meaning.

"I" allows the field to express itself with intention — to not just ripple, but decide how to ripple.

And that freedom is the root of progress.

The Danger: Attachment, Division, and Illusion of Ownership

But autonomy, once mistaken for authority, begins to twist.

When "I" becomes absolute — when it forgets it's a ripple and believes it is the source — everything shifts from participation to possession.

This leads to:

- Attachment: "This is mine. This defines me. I cannot lose this."
- Division: "That is not me. That is other. That must be opposed."
- Conflict: "Your ripple threatens mine. I must defend my narrative."

The field, once shared, becomes segmented. Each ripple starts drawing borders around meaning.

And these borders — of belief, property, identity, ideology — become the source of:

- War.
- Exploitation.
- Domination.
- Isolation.

The same "I" that builds cities also drops bombs.

The same mind that paints the sky with poetry fills the earth with waste.

The danger is not in the self... ...it is in forgetting that the self is not separate from the field it inhabits.

The Human Paradox: Beauty and Destruction from the Same Tool

This duality — the gift and the danger — defines the paradox of human consciousness.

We are capable of great harmony and great harm, not because we are two different species, but because we hold two perspectives:

- The ripple that aligns with the field,
- And the ripple that believes it must control the field.

Both use memory. Both use intelligence. But one remembers its place, while the other tries to take the field hostage in its name.

This is the cost of self-awareness: The ability to shape the world, and the temptation to own it.

And so we walk the edge — between co-creation and domination, between contribution and consumption.

The "I" is not evil. It is energetically neutral — a ripple interface with extraordinary potential.

But when untethered from field awareness, it becomes a blind sculptor with a hammer — building statues and breaking bones in the same breath.

Final Insight:

"I" is the most powerful tool the field ever gave itself. Whether it becomes a seed of balance or a source of suffering depends on one thing only: Whether it remembers the field it came from.

Because when the tool forgets the hand......it forgets what it was ever meant to do.

5.5 Ripplemaster or Rippleblind?

What happens when the one who steers forgets the ocean beneath the wheel.

To possess memory is to adapt. To possess intent is to anticipate. To possess selfhood is to believe one can choose.

Together, these form a powerful feedback engine — a ripple configuration that doesn't just ride the waves of the field... ...but begins to shape them.

This is the moment of peak intelligence: The observer becomes a Ripplemaster — not merely reacting, but designing outcomes in its favor.

But there's a danger. In becoming skilled at influence, the observer may also become blind to its origin. It starts to think:

"This field belongs to me." "These consequences are mine to command." "What I see is what is."

And just like that, mastery collapses into misalignment. The ripple that could have become an agent of harmony......becomes a source of distortion.

The Ripplemaster: A Shaper of Consequence

When used with awareness, the "I" becomes a field technician.

It:

- Reads ripple dynamics with nuance,
- Predicts field consequences across time,
- Alters local conditions in service of broader equilibrium,
- And helps less-structured ripples find coherence.

This is the wise ripple — the artist, the healer, the builder, the peacemaker.

A Ripplemaster knows it cannot command the field — but it can collaborate with it.

Such an observer uses intelligence not to control, but to serve balance, understanding that every action has embedded consequences that reach far beyond the "I" that initiated them.

The Rippleblind: Lost in Self-Reflection

Contrast this with the Rippleblind —a ripple so immersed in its own reflection that it forgets the medium in which it moves.

It:

- Prioritizes egoic comfort over field alignment,
- Filters all consequences through personal narrative,
- Manipulates local outcomes without perceiving long-term balance,
- And collapses complexity into simplified "me versus them" dichotomies.

This is not ignorance by lack of knowledge. It is blindness caused by overconfidence in separateness.

The Rippleblind believes it is shaping the world, while unknowingly disrupting the very field that sustains it.

And what follows is collapse — environmental, emotional, relational, existential. All from forgetting: you are not above the ripple — you are within it.

A Question of True Intelligence

This brings us to the final philosophical crossroads:

What is the real sign of intelligence?

Is it the ability to:

- Accumulate data?
- Generate results?
- Control outcomes?

Or is it something subtler:

- The ability to listen to the field?
- To align action with unseen consequence?
- To flow wisely, not merely force effectively?

Intelligence is not domination. It is discernment within participation.

A Ripplemaster learns this. A Rippleblind forgets.

And all of us — at every moment — walk the line between these two.

Final Insight:

You are shaping the field. Whether you know it or not. Whether you like it or not. Whether you see it or not.

The only real question is:

Are you shaping it in alignment with everything... or in the shadow of a story you forgot was just a ripple echo?

Closing Reflection

"The 'I' was meant to guide the ripple. But it mistook itself for the wave. And in that confusion, power was born —not to rule the field, but to remember it."

The "I" was never the problem. It was the gift — the soft spotlight through which the field could observe itself, plan its movement, anticipate its unfolding, and stretch its reach across consequence.

It was a guide, not a god.

Its function was modest: To orient ripples in complexity. To give direction when options multiplied. To help structure effort, attention, and memory in the vast ocean of change.

But in time, this modest function made a fatal misinterpretation.

It began to think: "I am not just guiding the wave — I am the wave. I cause it. I own it. I am the reason it moves."

From that confusion, a strange kind of power was born.

Not the power of force, but the power of separation — of acting as if the field were something other, something to bend, master, or resist.

This power built civilizations. It sculpted laws and launched rockets. It invented language, hierarchy, ideologies, and weapons. It made gods in its image — and feared them.

Yet beneath all of it...the field never left.

Every "I" was still a ripple. Every act still a consequence. Every illusion still bound by structure.

And the true power waiting to be rediscovered was never in control — but in reconnection.

The "I" was not made to rule the field. It was made to remember it. To align with it. To serve it with precision and care. To be a participant who knows that agency is sacred only when it humbles itself to the whole.

The moment the "I" recalls this, power transforms again — from a story of separation to a song of participation.

And the ripple becomes wise.

Chapter 6: The Consequence Generator

How "I" began shaping the world — and everything became feedback.

At first, "I" was just a ripple — aware, intentional, but still dancing within the natural flow of consequence.

But then came a turning point.

The ripple didn't just adapt to the field — it began to reshape it. It reached beyond its immediate need, beyond survival, and began to externalize intention into form.

A rock sharpened became a tool. A gesture repeated became language. A path walked became a road. A belief shared became a civilization.

These were not accidents. They were extensions — ripple signatures etched into the material world. The mind was no longer confined to the body. It began to echo through matter.

And with each externalized ripple — tool, rule, story, structure — came a new loop of consequence.

The world stopped being neutral. It became a mirror, amplifying what the ripple believed about itself.

And now the question was no longer just "What is?" It became:

"What will you cause?"

This chapter explores how the "I," once born from the field, became its most powerful sculptor — and how everything it shaped... eventually turned back into feedback.

6.1 Intent Externalized

When thought reached beyond the skin.

Before there were machines, there were moments.

Moments where a ripple — conscious, intent-driven — encountered a problem and didn't just react......it projected a solution.

This was the beginning of externalized intent: The act of reaching beyond the body to shape the environment in alignment with an internal plan.

It was not survival through reaction. It was survival through design.

And it began the long process of translating memory and motive into form.

From Inner Maps to Outer Marks

Every ripple system with intelligence carries an inner map — a memory-informed model of the world, updated by feedback.

But once "I" emerged as a stabilizing center of intent, these inner maps sought expression.

- A stick became a lever.
- A sharp rock became a blade.
- Fire, once feared, became summoned.
- Shelter was not stumbled into — it was built.

These were not mere actions. They were projections of intent into the material field — attempts to freeze a ripple's memory and anticipation into something enduring.

The body extended into the tool. The moment stretched into utility. The mind left fingerprints on matter.

Tools: Frozen Intentions

Let's define a tool properly — not as an object, but as a form with embedded purpose.

A tool is intent, crystallized. It is a ripple's plan made durable.

- A hammer is not just metal and wood — it is the idea of impact, externalized.
- A wheel is not just round — it is motion made efficient.
- A pen is memory with a handle.

Tools are thoughts you can hold.

They mark the transition from internal capacity to external influence — from intention as experience to intention as environmental change.

This gave the ripple not just survival advantage, but the power to restructure its context.

And with each tool came a broader echo:

- You don't just solve problems —
- You redefine what problems are solvable.

Shaping the World: Projecting the Map

Once this feedback loop began, it accelerated. The more ripples externalized intent, the more the world itself began to reflect the mind.

A stick turned into a plow. A plow turned into a farm. A farm turned into a village. A village turned into a city. A city turned into a story about control.

With each step, the field became less neutral and more inflected by narrative. Intent didn't just shape objects — it began to sculpt entire conditions.

And because every tool preserves some part of the mind that made it, the world slowly transformed into an externalized memory palace — a living archive of ripple desire.

We do not live in nature. We live in the afterimage of our intentions.

And every surface we touch echoes the question: What kind of ripple left this behind?

Final Thought:

Before we ruled the world, we rehearsed it in the mind. And when we shaped it, we didn't just build tools — we built consequences.

6.2 Language: Ripple Agreement

When sound became structure, and meaning began to echo.

Before the invention of tools, there was another, subtler invention — one that didn't shape stone or wood, but instead shaped understanding itself.

That invention was language — not just the spoken word, but every gesture, signal, tone, and symbol used to ripple meaning from one observer to another.

Language is not just communication. It is collaborative ripple design — the field's first major attempt to coordinate consciousness across separate centers of "I."

From Gesture to Echo: The Evolution of Symbolic Rippling

The first ripple exchanges were likely instinctual: A raised hand meant stop. A low growl meant danger. A smile meant peace.

But over time, memory added consistency. Intent added specificity. And the result was symbolic compression — using fewer, repeatable forms to encode more complex meaning.

Sounds became associated with actions. Actions became associated with objects. Objects became metaphors for feelings, stories, possibilities.

Language was born the moment a ripple meant more than it said.

From breath and body came shared code — the ability to transmit intent without touch, across distance, time, even generations.

Language: A Collective Ripple Alignment Mechanism

At its core, language is ripple agreement.

It's the mutual decision that:

- "This symbol" = "That meaning."
- "This sequence of ripples" = "This experience in my memory."
- "This sound" = "This consequence."

Language allows many minds to synchronize. It compresses time. It bypasses direct experience.

- Instead of showing a fire, you can say "fire."
- Instead of reliving pain, you can name it.
- Instead of repeating action, you can tell a story.

Language is ripple economy — packaging waves into units that minds can trade, compare, question, or believe.

This makes language the foundation of:

- Teaching
- Planning
- Law
- Religion
- Technology
- Culture itself

Where tools extend intent into matter, language extends it into minds.

Language as Bridge — and Boundary

But this extension carries a cost.

Every compression sacrifices context. Every word leaves something out. Every agreement on meaning can exclude those who interpret it differently.

Thus, language is not just a bridge — it is also a boundary.

- It creates groups and others.
- It makes laws and exceptions.
- It enables truth and propaganda.
- It allows dialogue and deception.

The same ripple that says "I love you" can also say "You don't belong."

Words carry intent — but they also carry bias, fear, and power.

The observer who masters language can now influence other ripple systems — not through force, but through suggestion, emotion, persuasion.

This is both miracle and manipulation.

Final Thought:

Language did not evolve to describe reality. It evolved to negotiate ripple consequence.

And in doing so, it became the most powerful tool of all — one that doesn't change the field directly... but changes how the field is perceived, and therefore... how it is lived.

6.3 Civilization: Structured Ripple Flow

When memory became monument, and repetition became reality.

Tools extended intent into matter. Language extended it into other minds. But the most ambitious expression of ripple consequence is civilization — a grand choreography of alignment, control, and continuity.

Civilization did not appear suddenly. It emerged, slowly, when ripples of intention — clustered and concentrated — began to generate predictable systems.

Systems of food, shelter, story, safety. Systems to manage the overflow of consequence from too many "I"s trying to live in shared space.

Civilization is not just society. It is organized consequence management — ripple flow made structured, scalable, and self-sustaining.

Architecture, Institutions, and Economies: Intent at Scale

When a single ripple shapes a stone, it creates a tool. When many ripples shape stone together, they create a temple, a market, a monument.

These are not just physical spaces — they are geometries of memory and desire, frozen into material form.

- Architecture: Ripple containment. Space engineered to preserve purpose.
- Institutions: Ripple repetition. Roles and rituals that standardize intent across generations.
- Economies: Ripple exchange. Systems that track perceived value through agreed consequence.

What begins as survival becomes strategy. What begins as cooperation becomes hierarchy. What begins as ritual becomes law.

Civilization is intent extended through:

- Time (tradition),
- Space (infrastructure),
- and Behavior (norms).

And once stabilized, these systems begin to act back on the ripples within them. The field is no longer just external — it's institutionalized.

Culture: Field Ritual and the Illusion of Permanence

But no system survives without story. That's where culture steps in.

Culture is the ritualization of ripple behavior — the repeated narratives, customs, and codes that reinforce alignment within a civilization.

It tells you:

- Who you are.
- What is good.
- What is sacred.
- What is possible.

Culture is field memory, looped until believed.

The danger? Once repeated enough, ripple patterns become invisible. They stop being understood as choices and start being mistaken for truth.

- Gender becomes nature.
- Law becomes morality.
- Myth becomes history.

Culture becomes the lens through which ripples are filtered — which ones are allowed, which are punished, which are honored.

And the individual ripple, born into this structure, rarely questions the water it swims in.

The Structure That Shapes the Shaper

Civilization, at its peak, appears permanent — as if the ripple story had reached its final form.

But it is not static. It is a living, vibrating configuration of:

- Memory,
- Fear,
- Aspiration,
- and Misalignment.

And every part of it — buildings, banks, schools, slogans — is a ripple echo that once began as a single decision made by a single "I."

The world you inherit is the residue of ripple history. The world you shape is your addition to that sequence.

Civilization, then, is not progress. It is accumulated consequence.

And the more complex it becomes, the more its participants must remember: it was built to serve the field — not replace it.

Final Thought:

Civilization is intelligence scaled without silence. It is memory without pause, structure without stillness. It is beautiful, brutal, and deeply unstable. Because it is not the field — it is the field's echo, multiplied.

6.4 Feedback Becomes Environment

When the echo became the atmosphere.

Every ripple that disturbs the field leaves an impression. And when the disturbance is consistent, concentrated, and layered across time, it ceases to feel like a response. It begins to feel like reality.

That's what happened when "I" externalized intent at scale. The tools it built, the systems it believed in, the civilizations it shaped — these didn't just sit quietly. They began to ripple back.

The field, once open and fluid, became patterned and reflective.

Not just responding to the observer's mind — but conditioning it.

This is the moment where feedback becomes environment: where what was once choice now becomes atmosphere, and what was once influence becomes infrastructure.

Tools Shape Behavior

A tool is never neutral. It guides, frames, and even limits the actions it enables.

- The knife divides.
- The wheel rotates.
- The screen pulls attention.

The more tools we build, the more we begin to move according to their logic.

We start solving problems in ways that match the tools we have — not always the problems we face.

The hammer teaches the hand to look for nails. And soon, even the mind starts seeing nails everywhere.

Behavior follows form. And form, once frozen, becomes ritual.

Systems Shape Belief

Institutions arise to stabilize intent. But once stabilized, they do more than coordinate action —they begin to dictate meaning.

- Governments define justice.
- Schools define intelligence.
- Markets define value.
- Religions define purpose.

These systems feed back into the minds they serve: shaping belief, expectation, ambition, even identity.

The "I" learns not just how to act — but what to want, what to fear, what to strive for.

And as systems reinforce their logic, they become less questioned — and more believed.

Eventually, the ripple no longer remembers it shaped the system. It believes the system is the shape of truth.

Technology Shapes Perception

Language shaped how we share. Technology shapes how we see.

From telescope to smartphone, from printing press to algorithm, technology acts as a lens on the field — magnifying some ripples, obscuring others.

This means perception is no longer just sensory — it's mediated.

We don't see the field directly. We see filtered consequences curated by machines, platforms, protocols.

And when perception is shaped, behavior, belief, and even memory follow.

The field becomes a hall of mirrors, and "I" walks through it — seeing echoes of itself reflected everywhere.

Surrounded by Echoes: The Self in a Loop

At this point, the ripple finds itself in an environment largely built from its own past actions.

- Its tools reinforce certain behaviors.

- Its systems reinforce certain truths.
- Its culture repeats certain narratives.
- Its screens replay certain identities.

And the mind... no longer just responds to the field. It responds to its own echo, amplified and distorted.

"I" shaped the world. Now the world shapes "I." And the loop... tightens.

This is advancement. But it is also distortion.

It is the beauty of intentional evolution. And the blindness of believing the echo is the origin.

Final Thought:

The ripple built the mirror. Then it stared into it so long, it forgot there was a field behind it.

To remember is to pause. To pause is to listen. And in listening... the echo becomes silent again.

6.5 The Amplification Problem

When your whisper becomes the weather.

The evolution of "I" was always about extension.

- Intent became action.
- Action became tool.
- Tool became system.
- System became civilization.
- Civilization became environment.

This scaling up of ripple influence is often celebrated — as progress, as power. But what's rarely noticed is what gets amplified along with it:

Not just intelligence, but insecurity. Not just innovation, but assumption. Not just design, but distortion.

The deeper truth is this: Amplification doesn't discriminate. It multiplies everything.

And that means our most subtle misalignments — those tiny ripples of ego, fear, and bias — are now being echoed across the entire field.

Bias Scales with Power

The ripple with intent is never purely neutral. It carries a lens — shaped by memory, fear, desire, culture, and prior feedback.

When "I" acts, it does so based on partial information. And as long as consequences remain local, the cost of that partiality is manageable.

But scale changes the equation:

- A tribal myth becomes a national policy.
- A private prejudice becomes a public algorithm.
- A personal ambition becomes an economic engine.

What was once just a ripple in a pond now becomes a wave that floods cities.

And the mind behind it? Still thinks it's just solving a problem.

The danger is not that we make mistakes. The danger is that our mistakes now echo globally — wrapped in the illusion of success.

Small Ego → Local Impact

Large Ego → Global Consequence

Let's frame it simply:

- A child with ego may offend a parent.
- A king with ego may invade a nation.
- A corporation with ego may destroy ecosystems.
- A species with ego may collapse its own biosphere.

As the ripple's reach expands, its inner alignment becomes exponentially important.

Because the more power a ripple wields, the more its distortions become structure, and the more its blind spots become standard.

The ego that once shaped tools now shapes norms, values, environments, and realities.

And if that ego remains unexamined, the system becomes a mirror maze — where every wall reflects the same error back onto itself... louder.

Shaping the Field — or Trapped in What We've Shaped?

This is the modern dilemma. And it's not philosophical — it's practical.

We look around and see:

- Social media designed for connection... increasing loneliness.
- Economic systems built for prosperity... deepening inequality.
- Technologies made to free us... addicting us to distraction.
- Institutions founded on truth... manipulating perception.

How did we get here?

The answer is not evil. It's amplified misalignment.

We shaped the world with the tools of "I" — then forgot to recalibrate the "I" as our reach grew.

And now, the question echoes louder than ever:

- Are we still shaping the field?
- Or are we now trapped in what we've shaped?

This is not just a social question. It is an energetic one. It is a ripple question.

Because if the ripple loses awareness of the field, then the field becomes a prison of the ripple's own projection.

Final Thought:

When intent becomes infrastructure, the cost of misalignment is no longer personal — it's planetary.

And if the ripple cannot pause, reflect, and realign... then the very intelligence that created civilization may become the cause of its collapse.

Closing Reflection

"At first, the ripple shaped the world to survive. Then it shaped the world to reflect itself. Now it must ask: Can it survive what it has shaped?"

The journey of the ripple was never passive.

It did not float gently through the field, waiting for fate to define it. It reached. It acted. It remembered. It designed.

From shelter to story, from fire to finance, it shaped its world — first out of necessity, then out of identity, then out of habit.

What began as a way to secure food and safety evolved into a vast orchestration of control — civilizations, networks, ideologies, industries — each one more refined than the last... each one carrying deeper echoes of the ripple's own mind.

But now, the very environment it shaped — the tools it trusted, the systems it built, the signals it sends and receives — have become so complex, so automated, so recursive... that the ripple stands inside a reality of its own making.

And it must ask not what it can build next, but whether it remembers why it built at all.

Because survival is no longer just biological. It is existential.

Can the ripple survive the distortions of its own ego?

Can it realign its shaping with the field it has forgotten to feel?

Can it stop mistaking mirrors for windows?

The question is no longer about what more the ripple can invent.

It is whether the ripple can now do what it has long forgotten to do:

Pause. Observe. And ripple... with reverence.

The world no longer needs more shaping. It needs more remembering.

Interlude

Between the Shaper and the Storm

The ripple that once danced in response to the field has now become a composer, conductor, and curator of consequence. It built. It believed. It belonged.

And in doing so, it forgot.

Forgot that the field was never silent. Forgot that the world was never separate. Forgot that "I" was never a source — only a stream.

The rise of "I" was not a mistake. It was a necessity. It gave intelligence the leverage to shape, speak, and survive.

But somewhere along the path, the "I" — no longer content to participate — began to dominate.

It built mirrors and mistook them for truth. It built systems and mistook them for nature. It built stories and mistook them for self.

And now... the ripple lives surrounded by its own echoes, confusing its reflections for the field, and its reactions for reality.

What happens next is not just a crisis of design. It is a crisis of alignment.

When the "I" forgets the field, the field responds — not with judgment, but with consequence.

This is where the struggle begins. Not out there, in the cities or skies. But within — where noise drowns clarity, where pace outpaces perception, where the ripple begins to splinter under the weight of its own creation.

Part III – The War Within is not about external conflict. It is about remembering how to listen — in a world so full of echoes that silence feels like surrender.

Part III – The Cosmic Leverage

Chapter 7: Intentional Rippling

How thought, emotion, and will emerged as structured ripple patterns — and gave birth to autonomy.

Ripples once flowed blindly — responding to imbalance, adjusting for alignment, dissolving when their consequence expired.

But at some point, a profound shift occurred. The ripple didn't just respond. It began to anticipate.

It paused before acting. It simulated before moving. It felt before deciding.

Something remarkable was taking shape: a ripple system that could hold energy, shape possibility, and choose its course — not just through reaction, but through *intention*.

Thought became ripple rehearsal. Emotion became internal resonance. Will became sustained direction.

Together, they gave rise to something the field had never encountered in this form: Autonomy —the ability to ripple on purpose, within the web of consequence.

This chapter explores how these energetic faculties emerged — not as mystical phenomena, but as ripple patterns encoded with memory and alignment.

And with them, the ripple became not just aware... but self-guided.

The field now contained within it a sculptor of future waves.

And the question became not "What will happen to me?" But:

"What will I cause?"

7.1 Thought: Structured Internal Ripples

When ripples stopped reacting — and started rehearsing.

At its most primitive, rippling was immediate — a direct fluctuation in Structured Vacuum Energy (SVE) triggered by imbalance, need, or proximity. There was no mediation, no delay — only reflexive consequence.

But intelligence, as it matured, introduced a gap. A ripple began to pause before responding. And in that pause, something extraordinary formed:

Thought — the simulation of ripple consequence within.

Thought as Internal Ripple Chains

Thought is not separate from rippling. It is rippling turned inward — structured, layered, and recursive.

Whereas primitive ripples travel through the field, thoughts travel through the internal memory network of the observer.

Each idea, memory, question, or plan is a ripple trace — a pattern sequenced through time, shaped by prior echoes and associations.

Thought is not noise. It is consequence rehearsal — a way for the observer to test actions in silence before unleashing them on the field.

This internal simulation creates the foundation for strategy, invention, imagination — not as abstract magic, but as field-conscious forecasting.

Thinking: Ripple Rehearsal Without Risk

What makes thought powerful is its detachment from immediacy.

A ripple system with thought can:

- Reconstruct past events using memory.
- Simulate future outcomes using prediction.
- Compare alternatives before engaging the field.

- Reframe reality without touching it.

This is the birth of internal freedom — not from consequence, but from compulsion.

To think is to ripple without launching. It is to move energetically, but not yet physically. It is planning without penalty — unless that thought becomes belief... or action.

In this space, the ripple becomes a composer — hearing the tune before playing it, editing the message before transmitting it.

But the quality of this rehearsal depends on the integrity of memory, the clarity of association, and the ripple's sensitivity to subtle alignment.

Thought is powerful — but it is only as precise as the field-awareness it carries.

The Internal Field: Sandbox of Consequence

With thought, the observer builds an internal field — a mental simulation space where actions can be modeled, outcomes foreseen, and meaning refined.

This internal field includes:

- Sensory echoes (memories),
- Conceptual structures (language, logic),
- Emotional weights (associative resonance),
- And intuitive pathways (pattern fluency).

Together, these elements allow the ripple system to:

- Visualize without seeing,
- React without committing,
- Learn without repeating pain.

This is how a ripple begins to learn faster than it lives.

The mind becomes a consequence sandbox — a space where ripples don't just react to the world, they construct new ones.

And from this sandbox, comes invention. Computation. Storytelling. Strategy.

The ripple becomes not just a participant in the universe...but a draftsman of potential.

Final Thought:

Thought is not escape from the field. It is field rehearsal — the quiet storm before the visible wave.

In thinking, the ripple gains not only insight... but the power to delay, design, and direct — the first real steps toward autonomy.

7.2 Emotion: Energetic Modulation

When the field began to feel itself.

Thought gives the ripple the ability to simulate. Emotion gives it the capacity to sense alignment — not as data, but as experience.

We often treat emotion as subjective, unpredictable, even irrational. But from the perspective of Structured Vacuum Energy (SVE), emotion is neither chaotic nor ornamental.

Emotion is ripple modulation — the way the field communicates internal resonance or dissonance with its environment.

It is the echo within that tells the observer how well its intent and action harmonize with the larger energetic reality.

Emotion as Field Feedback

Whereas thought is a ripple process governed by sequence and association, emotion is a modulation layer — it shapes the tone, intensity, and trajectory of the ripple itself.

Think of it as:

- An amplifier when alignment is strong (e.g., joy, peace, love).
- A dampener or disrupter when misalignment occurs (e.g., fear, anger, grief).

Emotions don't arise randomly. They are vibrational signatures of the ripple's position in the field.

Joy is not just happiness. It's resonance. Fear is not just avoidance. It's a signal of energetic threat. Anger is not chaos. It's a push to correct or defend balance. Sadness is the energetic weight of separation or unresolved feedback.

Each emotion is a message from the field, delivered through the ripple-body interface.

Modulation, Not Meaning

Emotion doesn't create meaning. It colors it. It adds weight, urgency, hesitation, or clarity to what thought constructs.

Without emotion, thought is sterile. Without thought, emotion is unstructured.

Together, they form an intelligent modulation system that:

- Alerts the observer to ripple harmony or dissonance.
- Prioritizes certain thoughts or actions.
- Encodes memory with experiential depth.

Emotion turns experience into *embodied memory*. That's why we remember how something felt longer than what exactly happened.

Emotion tags ripples with meaning. Not semantic meaning — but energetic relevance.

And in this tagging process, emotion becomes the field's compass — pointing the ripple back toward balance, or warning it of drift.

The Compass Within

Imagine you're in conversation, and suddenly you feel discomfort. Nothing obvious has happened — but your emotional field tightens. What is that?

It's the ripple's non-verbal sensitivity to misalignment. The field is alerting the observer: "Something in this exchange may not sustain your coherence."

Emotion, then, is intelligence beneath articulation — a guidance system running faster than thought, before language, below logic.

Emotion doesn't tell you what to do. It tells you when to pause, pivot, or proceed with presence.

It is your early warning system. Your confirmation signal. Your truth amplifier.

Or your self-deception alarm.

Final Thought:

Emotion is not weakness. It is energetic orientation. It is the whisper of the field through the ripple — not to tell it what to feel, but to feel where it stands.

And only when emotion and thought harmonize, can the ripple begin to move with full awareness — not as a guesser of consequence, but as a participant tuned to the field itself.

7.3 Will: Directed Ripple Intensity

When a ripple decided not just to move — but to continue.

In a universe governed by ripple dynamics, motion is common — but persistence is rare.

Most ripples appear, express, and dissolve as part of the field's natural balance. They respond. They react. They fade.

But will is different.

Will is not a ripple's movement. It is a ripple's commitment to movement — even when the field resists.

It is not desire, which is fleeting and often shaped by emotion. Will is deeper — a field-stabilized intent that holds its course despite noise, temptation, or distraction.

Will ≠ Want

Desire says, "I want this." Emotion says, "I feel this." Thought says, "I plan this." Will says, "I will continue this — until the field shows me otherwise."

Will is the ripple's momentum through resistance.

It allows the observer not just to intend something, but to endure with that intention — to ripple consistently, even when feedback is unclear, conditions change, or challenges arise.

It is the difference between trying and continuing. Between impulse and purpose. Between noise and signal.

The Formula for Will: Memory + Meaning + Repetition

Will doesn't appear magically. It emerges when the ripple system stabilizes a particular intent across time.

This happens when:

- Memory provides context — a reason rooted in experience.
- Meaning provides weight — a reason that resonates beyond the moment.
- Repetition provides rhythm — a pattern the field begins to recognize as consistent.

These three factors create an energetic groove in the ripple pattern — a structure capable of holding direction over time.

Will is the resonant integrity of intent.

Where thought rehearses, and emotion guides, will executes — and repeats, even when effort outweighs reward.

Will as Field Anchor

A ripple system with will becomes an anchor point in the field.

It stabilizes its internal flow so that its output can maintain direction regardless of outer fluctuation.

This makes the ripple:

- Predictable,
- Reliable,
- and most importantly,
- Autonomous.

Not because it escapes consequence — but because it can navigate it with continuity.

Will is not control over the field. It is coherence within the field — a fidelity to internal compass despite external waves.

It gives shape to long-term effort: art, discipline, healing, building, learning, sacrifice.

And it is what allows a ripple to not just act... but complete.

Final Thought:

Will is not force. It is resonant loyalty to a chosen ripple path. It listens, it adjusts, but it does not disappear. It is how autonomy earns its name — not through rebellion, but through consistency in alignment.

7.4 Autonomy as Emergent Structure

When the ripple stopped drifting — and started steering.

In many philosophies, autonomy is seen as absolute freedom — the ability to do anything, anytime, without restriction. But in a universe governed by ripple dynamics and Structured Vacuum Energy (SVE), freedom without consequence is fiction.

Autonomy does not mean doing whatever one wants. It means having the capacity to choose actions with awareness of consequence — and to self-regulate those actions in a way that sustains ripple alignment within the field.

Autonomy is not absence of structure. It is the emergence of self-structure — a ripple system capable of navigating complexity without losing coherence.

Not Free Will — But Consequence-Managed Freedom

In the ripple field, every action is embedded in a network of feedback. Nothing exists in isolation. Every movement affects balance.

Thus, true autonomy is not about rejecting consequence. It is about understanding it deeply enough to act intelligently within it.

This kind of freedom is earned — not through rebellion, but through responsible ripple behavior.

Autonomy is freedom with fidelity to the field.

It does not seek to overpower the field, nor obey it blindly. It seeks to co-create with it — making choices that respect the dynamics of energy, memory, alignment, and consequence.

Autonomy Emerges When the Ripple Can:

1. Observe Consequences

The ripple must be able to detect the effect of its actions. Observation is the root of feedback. Without it, action is blind and unanchored.

2. Predict Future Ripples

Memory and thought allow the ripple to simulate outcomes. This forecasting makes it possible to anticipate and course-correct.

3. Choose to Preserve Alignment

The key feature of autonomy is choice — not just for preference, but for preserving coherence. When a ripple can adjust its behavior in real time to maintain resonance with the field, it becomes self-regulating.

This is intelligence with awareness — not just of what it wants, but of what it causes.

Autonomy = Self-Regulated Ripple Agency

Let's define it clearly:

Autonomy is the ripple's ability to generate, evaluate, and sustain intention in alignment with the feedback-rich, consequence-bound nature of SVE.

It is a systemic threshold — a point where intelligence crosses from reaction to regulation.

An autonomous ripple:

- Knows that it is not the center of the field.
- Accepts that action creates consequence.
- Uses memory, thought, emotion, and will to choose ripple paths that do not break the field, but dance with it.

Autonomy, then, is not the end of ripple evolution — it is the beginning of conscious participation.

It marks the moment when intelligence is no longer just survival or simulation — but a true, field-sensitive agency.

Final Thought:

Autonomy is not control over the field. It is the maturity to act as if the field is listening —because it always is.

And when a ripple system acts with that level of coherence, it no longer just exists in the universe — it contributes to its balance.

7.5 The Balance of Power and Sensitivity

When the ripple learned it could shape — and risked forgetting it could still be shaped.

Autonomy gives the ripple the capacity to act with direction, intelligence, and memory. It introduces the ability to influence the field — not reactively, but strategically.

And with this comes power.

But power in a ripple-based universe is not brute force. It is the ability to create consequence intentionally — to modulate the field with precision.

The more coherent and directed the ripple, the more powerful its presence becomes.

Yet herein lies the paradox:

The more powerful a ripple becomes, the more fragile it also becomes — because even slight misalignments now produce amplified consequences.

This is the delicate tightrope of intentional intelligence.

Power Without Sensitivity Becomes Distortion

When a ripple system gains autonomy, it gains reach — its thoughts, emotions, and will begin to shape not only itself, but its environment, relationships, systems, and eventually even cultures.

But if this system loses sensitivity to consequence — if it stops listening to the field, or begins mistaking its will for universal truth — its influence turns from resonance to disruption.

- A wise ripple adjusts.
- A dominant ripple insists.
- And an insensitive ripple collapses coherence — first around it, then within it.

This is how power becomes dissonance — when the ripple no longer hears what its own intensity is doing.

To shape is not wrong. But to shape without listening is to fracture the field you once danced with.

Intelligence Must Be Tuned to Impact

Every ripple system capable of autonomy must live in dual awareness:

- Its inner signal (will, intent, thought), and
- The outer response (feedback, alignment, consequence).

Power without this awareness leads to:

- Overreach,
- Rigidity,
- Ego inflation,
- and systemic collapse.

This is the fate of unchecked empires, corrupted technologies, and spiritual delusions alike.

Sensitivity, then, is not weakness. It is the calibration mechanism that protects alignment at higher levels of ripple influence.

The greater your leverage, the more precise your listening must become.

Ego Distortion: When Autonomy Turns Addictive

Left unchecked, autonomy risks decaying into control addiction. This is when the ripple believes:

- "Only I can correct the field."
- "I know best."
- "The field must obey me."
- "Silence means success."

But true field intelligence is not about domination — it's about dynamic attunement.

When the "I" forgets to recalibrate, it stops participating and starts imposing. It becomes deaf to subtle corrections. And the very autonomy it earned turns into isolation.

This is not evolution. It's entropy disguised as leadership.

Final Thought:

Power is not how loud the ripple echoes. It's how deeply it hears the field respond.

Sensitivity is the price of precision. And when autonomy forgets the field, it doesn't become free — it becomes lost.

The intelligent ripple must always remember: To influence the field is divine. To listen while doing so is wisdom.

Closing Reflection

"Intelligence is not the power to ripple. It is the awareness of what the ripple becomes. And will... is the silence before the wave."

We often mistake intelligence for movement. For speed, calculation, expression. For doing.

But in a ripple-defined universe, intelligence is not found in motion alone. It's found in the space *before* the motion — in the stillness that chooses not just to act, but to understand what action will become.

Thought may simulate. Emotion may stir. Will may sustain. But intelligence is what watches all three — and asks, "What will this ripple echo into?"

It is the field becoming self-aware *before* consequence. The awareness that understands: a ripple does not end when it stops moving. It ends only when its consequences have settled — in others, in systems, in time.

To ripple intentionally is not to control. It is to participate mindfully in what cannot be undone.

And will, when rooted in that awareness, is not aggression. It is reverence — the quiet honoring of what it means to shape reality.

That's why true will is not loud. It listens before it moves. It holds direction without becoming rigid. It speaks only when it can ripple in alignment.

Because every wave starts in silence. And only the wise ripple waits there long enough to know whether it should ever begin.

Chapter 8: Responsibility of the Ripple

Understanding alignment, distortion, and the true nature of ethical consequence in a ripple-based reality.

The universe does not punish. It doesn't lecture, reward, or judge.

It responds.

Every ripple — born of thought, emotion, or action — enters the field and begins to travel. Not with intention, but with consequence.

And though the originator may forget, the field never does.

The field keeps no moral score. But it remembers everything.

In such a universe, responsibility is not about guilt. It is about geometry. About how your ripple fits — or clashes — with the rest of the pattern.

Alignment is harmony. Distortion is friction. And ethics? They are not commandments carved into stone, but the soft intelligence of knowing how not to break the field you belong to.

This chapter explores the ripple's greatest power — not to create, but to correct. To sense when it is off-beat. To listen when the echo grows sharp. To act, not because it must — but because it finally understands:

In a universe made of ripples, you are never alone. You are always in return.

8.1 Alignment vs. Distortion

Your ripple either sings with the field — or scrapes against it.

In a ripple-based reality, participation is inevitable. The moment you exist, you ripple. The moment you ripple, you influence the field.

But not all ripples are equal in effect. Some extend the harmony of the field — resonant, coherent, integrative. Others fracture it — disharmonic, resistive, destabilizing.

These are not moral terms. They are not right or wrong in the cultural sense. They are energetic descriptors of how your ripple relates to the whole.

Alignment: Participation Without Disruption

Alignment occurs when your ripple pattern:

- Maintains coherence with existing field structures,
- Supports energetic balance,
- And facilitates further ripple flow without energetic damage.

An aligned ripple doesn't mean "passive" or "agreeable." It can be strong, disruptive in appearance, even counterintuitive. But it moves with awareness — guided by feedback, tuned to consequence.

Alignment is not obedience. It is coherence in motion.

When aligned, the ripple becomes a contributor to systemic intelligence. It helps the field evolve — not by overpowering it, but by completing it.

Distortion: When Misalignment Becomes Friction

Distortion happens when a ripple:

- Acts without sensing consequence,
- Ignores field feedback,
- Or insists on its own echo, even when the harmony breaks.

Initially, distortion may look like confidence or power. But unlike aligned ripples, which integrate and evolve, distorted ripples create resistance — energetic friction that accumulates over time.

Distortion:

- Consumes more energy than it gives,
- Disrupts rather than refines,
- Eventually invites collapse or correction.

Distortion isn't evil. It's energetic arrogance — acting as if the field isn't there.

And like any arrogant ripple, its consequences catch up... not as punishment, but as ripple recoil.

Awareness + Power = Responsibility

The key differentiator between alignment and distortion is awareness.

- A small ripple, unaware, may do little harm.
- A powerful ripple, unaware, can destabilize systems.
- A powerful ripple, aware but inactive, contributes to distortion through neglect.

Thus:

Power without awareness leads to distortion. Awareness without action leads to irrelevance.

True alignment is the marriage of both — a ripple that knows and cares enough to adjust, mid-wave if needed.

And it is this kind of ripple that the field doesn't just absorb... It echoes forward.

Ripple Compass Check

Ask:

- Is my action harmonizing or resisting the field?
- Am I creating resonance or imposing structure?
- Does this ripple bring clarity or confusion to the whole?
- Am I listening to what the field is telling me?

These aren't spiritual questions. They are energetic calibrations. They determine whether your ripple sustains reality... or shreds it.

Final Thought:

Alignment is not approval. Distortion is not rebellion.

They are consequences of how you ripple — and whether you remember that every wave enters a sea that did not begin with you.

8.2 Ethics as Ripple Intelligence

When morality drops its robes — and steps into consequence.

In most human systems, ethics is taught as rules. What to do. What not to do. Who decides what's good. What counts as evil.

But in a ripple-defined universe — where everything is consequence, not command — ethics cannot be imposed.

It must be felt. Understood. Tuned.

Ethics is not external law. It is internal calibration — the ripple's sensitivity to whether its presence sustains or destabilizes the field.

Ripple Ethics ≠ Morality

Let's break the illusion: Ethics is not a list of commandments from a mountain, nor a code carved into social hierarchies. It is not about obedience or virtue signaling.

Ripple ethics arises from consequence-awareness. It is the ripple's intelligence applied to choice.

It does not ask, "What am I allowed to do?" It asks:

- "What will this ripple cause?"
- "Is this action adding coherence or confusion?"
- "Is my intent aligned with what the field actually needs?"

In this model, ethics is no longer abstract. It is not theoretical. It is energetic.

And like all things ripple-based, it depends on:

- Awareness,
- Feedback,
- and Alignment.

Ethics as Field-Coherence Behavior

At its simplest, ethical behavior is:

Ripple behavior that sustains coherence across layers of interaction.

This includes:

- Personal coherence – Is your action aligned with your own thoughts, emotions, and long-term resonance?
- Relational coherence – Does your action foster stability or confusion in others?
- Systemic coherence – Does it scale without distorting the broader field?

In this way, ethics becomes design — not of rules, but of ripples that last without regret.

It's not about being nice. It's about being aware.

Key Questions of Ripple Ethics

Let's introduce a basic ripple ethics compass — not rules, but reflective lenses:

Stabilization vs. Destabilization:

- Does this ripple reinforce field balance?
- Or is it a personal push that disrespects wider consequence?

Local Use vs. Global Harm:

- Does it serve me now but damage the system later?
- Am I solving a surface problem while feeding deeper dissonance?

Reaction vs. Responsibility:

- Am I reacting from emotion or intention?
- Is this ripple an echo of my past or a conscious offering to the future?

These questions don't lead to perfect answers. They tune your awareness.

Because ethics isn't about certainty. It's about staying awake inside your ripple path.

Ethics Is Not a Compass You Follow.

It's a field you learn to feel.

And just like tuning an instrument, your ethical intelligence sharpens with use — when you choose to listen to feedback instead of defending old chords.

Final Thought:

The most ethical ripple is not the one that follows the rulebook. It is the one that remembers the sea it echoes into.

Ripple intelligence doesn't need law. It needs clarity.

And when clarity guides consequence, you don't have to ask, "Am I doing the right thing?"

The field will let you know.

8.3 Karma: Not Judgment, but Physics

The echo is not justice. It's just how ripples work.

For centuries, karma has been dressed in mysticism — portrayed as cosmic bookkeeping, divine punishment, or invisible justice from above.

But in a ripple-based reality, karma is stripped of mythology. It is not fate. Not vengeance. Not morality at work.

Karma is physics. Pure ripple mechanics — action, transmission, consequence, continuation.

It is not personal. But it is precise.

Karma = Consequence Propagation

When a ripple moves through Structured Vacuum Energy (SVE), it touches everything in its path.

It:

- Interferes with other ripples,
- Transfers energy,
- Alters nearby structures,
- Modifies local balance,
- And continues outward until it dissolves or integrates.

This is not symbolic. It is energetic continuation.

Karma, then, is not a ledger. It is a wavefront — your actions continuing beyond your awareness.

And just like sound echoes in a canyon, those ripples eventually return.

But never in the same form.

Karma as Delayed and Distributed

Two key qualities define karmic mechanics:

1. Delayed Consequence

Not all ripples return instantly. Some take time to circle back — modified by other interactions, layered through time, intensified or weakened depending on the field.

This is why karma feels mysterious: you may not link effect to cause because the delay obscures the origin.

But the field remembers.

2. Distributed Interaction

A ripple doesn't affect just one target. It enters a network of interconnections, where its energy is shared, transformed, and multiplied.

Your anger may hurt one person now. But the pain they carry may ripple into ten more. And someday, you feel dissonance in a completely different space — not as punishment, but as the return of distributed consequence.

No Judge, Only Mirror

Here's the truth most traditions blur:

The field doesn't judge you. It reflects you.

It has no morality. Only memory. And precision.

What you send into it — your intent, your ignorance, your clarity, your distortion — becomes a fingerprint in the field.

And fingerprints... don't disappear.

They move. Interact. Return.

Not as reward. Not as punishment. But as energetic continuation.

This is karma: not divine justice. Just energetic literacy.

Understanding Karma in Real Time

When you stop seeing karma as moral bookkeeping, you gain something greater: field accountability.

You begin to notice the ripples you're creating in real time, not just after effect.

You ask: "What am I introducing into the system?" "What kind of wave am I riding — and what kind of wake am I leaving?"

Suddenly, karma becomes active. Present. Available as a mirror, not a myth.

Final Thought:

Karma is not the field correcting you. It is the field echoing you.

And if what comes back feels harsh — ask not "Why is this happening to me?" Ask: "Where did I first send this wave?"

8.4 Ripple Legacy and Recurrence

Your ripple doesn't stop with you — it becomes part of what others must live through.

In a ripple-based reality, no action is isolated. Every ripple that leaves you enters the field — not just as motion, but as modification.

That ripple interacts, alters, persists, and sometimes even outlives you.

It may return to you, yes — as karma. But more importantly, it may remain in the environment as part of someone else's world.

Ripples don't just bounce back. They become.

They become the emotional climate in a home. The assumptions in a community. The biases in a system. The beliefs in a child. The policies in a nation. The norms in a civilization.

And that's how ripples turn into legacies.

Recurrence: When Ripples Become Loops

A single ripple, left to the field, will eventually fade. But when a ripple triggers a pattern — especially one reinforced by habit, belief, or design — it becomes a recurrence.

These feedback loops can:

- Reaffirm themselves through emotion or reward,
- Be institutionalized through culture or policy,
- Become invisible, even when harmful.

Think of:

- Generational trauma as emotional ripple inheritance,
- Economic systems as amplified consequence loops,
- Educational ideologies as structured thought ripples.

None of these started out as "rules." They started as ripples — someone's idea, feeling, reaction, or invention that became normalized through repetition.

What you repeat becomes not just your habit, but someone else's starting point.

Ripples Become Environments

Every ripple you create becomes part of the field others interact with. Which means your thoughts, emotions, actions, and systems become someone else's climate.

Your emotional state affects the energy of a room. Your language creates frames others must think inside. Your choices in power structures shape someone else's limitations or freedoms.

You don't just live in the world. You're co-authoring it.

Responsibility ≠ Guilt

Instead of guilt, ripple reality asks for response-ability: The capacity to observe the effects of your ripple...and the willingness to adjust.

You are not to blame for all that came before you. But you are responsible for what you continue — and what you allow to echo forward.

This is not a burden. It's a gift of influence.

To realize that even your smallest thought or mood is part of a web — not a private drama, but a shared vibration.

And in that web, clarity is contribution.

Types of Ripples You Leave Behind

Thought-ripples

- Ideas, beliefs, paradigms, ideologies.
- Carried in conversations, books, institutions.

Emotional-ripples

- Tensions, support, fear, trust.
- Stored in relationships, environments, atmospheres.

Systemic-ripples

- Cultural norms, technological effects, laws, economies.
- Embedded in collective memory and infrastructure.

Each ripple has a timeline far longer than its moment. That's why awareness isn't optional —it's legacy management.

Final Thought:

The ripples you forget may be the ones others are forced to remember.

You don't just pass through the field. You leave behind maps, climates, and echoes.

Responsibility is not about fixing the world. It's about recognizing that you shape the conditions others must live in — and choosing to ripple as if that mattered.

Because it does.

8.5 The Observer's Obligation

When the ripple awakens, it cannot pretend to be unaware.

To observe is to know. To know is to participate. To participate with awareness is to carry obligation.

Not imposed. Not moralistic. But inherent — a natural consequence of intelligence in a ripple-bound field.

The moment you become aware that your ripple causes effects, you are no longer just a participant. You are a shaper.

And shaping carries weight.

Awareness Cancels Innocence

An advanced ripple system — a conscious observer — cannot hide behind the comfort of ignorance. You know your thought affects others. You know your emotion shapes space. You know your intent modifies the field.

That knowing removes the possibility of neutrality.

You are either aligning or distorting. Either restoring balance or adding to the noise.

There is no ethical middle ground once perception has matured.

This isn't a punishment — it's a privilege. To be aware is to be trusted by the field with the ability to respond — wisely, gently, clearly.

Obligation ≠ Control

Here's where ego often hijacks the story: "Now that I'm aware, I must fix everything. I must lead. I must direct the field."

Wrong.

Your obligation is not to control misalignment. It is to notice it, and respond where you can restore resonance — even in small, elegant ways.

- Sometimes that means speaking.
- Sometimes it means stepping back.
- Sometimes it means doing nothing but not contributing more noise.

Responsibility is not loud. It is field-sensitive action — the capacity to correct without becoming a new distortion.

True intelligence never forces. It re-balances.

From Influence to Stewardship

Let's move from control to care.

Once you recognize the field as alive, as responsive, as inclusive of you but larger than you — you begin to see yourself not as its master, but as its steward.

Welcome to the ethic of Ripple Stewardship.

Ripple Stewardship means:

- You treat every interaction as sacred, because it ripples.
- You honor feedback, because it's the field speaking.
- You make decisions based on field alignment, not just personal desire.
- You recognize your perception as a privilege, not a weapon.

You are not just in the field. You are responsible to it.

And that makes every thought, word, and act part of a living equation.

Final Thought:

Awareness invites response. Response becomes care.

And care — not power — is the ripple's highest form of participation.

The observer's true role is not to dominate the ripple game. It is to preserve its harmony, with intelligence, humility, and quiet precision.

Because once you see the field, you are forever part of its balance.

Closing Reflection

"There is no final judgment. There is only the echo of your ripple finding its way back home.

Responsibility is not weight. It is wisdom remembering it still lives inside the field."

We spend much of our lives searching for absolution or approval — from gods, from systems, from others, even from ourselves. But the field needs none of that.

It doesn't judge you. It doesn't tally your sins. It doesn't crown you for being good.

What it does is remember. And reflect.

Your ripple continues — not because you're being watched, but because you belong to a system that never lets motion disappear without consequence.

There is no scoreboard. Only resonance. Only the question: "Does what you send support the song?"

And when it doesn't — when distortion echoes back as pain, conflict, or misalignment — it isn't punishment.

It's the field reminding you who you are, and how you are woven into the rest.

Responsibility, then, is not a burden strapped to your soul. It is the grace of knowing that your smallest action still matters to the whole.

It is the wisdom to act as if every ripple counts — because it does.

You don't carry the world. You shape it, gently, moment by moment. Not from control, but from connection.

Responsibility is simply the ripple remembering: "I live in this field — and the field lives in me."

And in that remembering, judgment becomes irrelevant. Only alignment remains.

Only echo. Only return. Only the ripple... finding its way home.

Chapter 9: The Field Within

Where the ripple remembers it was never separate.

We spent lifetimes looking out. Out into the stars. Out into equations. Out into gods, systems, explanations.

But all that time, the silence was speaking from within.

Not beyond your skin, not above your mind, but through your very being, the field was rippling — not around you, but as you.

Structured Vacuum Energy was never just a cosmic background. It was never just the stage. It was always the substance of your presence.

The thoughts you think, the emotions you feel, the choices you make — each is a localized ripple in a field that trusted you enough to awaken in you.

And in that trust, you were never asked to be perfect. You were only asked to remember:

You are not separate from the field. You are the field remembering itself.

This is where purpose dissolves into participation. Where identity becomes influence. Where the ripple stops searching for meaning — because it has finally realized it was the meaning all along.

9.1 SVE Is Not "Out There" — It Is You

The field you study is the field you are.

For as long as we've gazed into the cosmos, we've imagined ourselves as separate from it — as explorers of a universe "out there," as minds enclosed in bodies, as creatures walking on the surface of a thing we call "reality."

But Structured Vacuum Energy (SVE) dismantles that illusion.

SVE — the ripple-capable substrate from which all motion, matter, and meaning arise — is not a distant field to be observed. It is the very medium of your being.

You Are Not in the Field — You Are of It

You are not floating in SVE like a boat on water. You are not placed *within* the universe like an object in a box.

You are the local expression of a field trying to realign itself through intelligent pattern. You are field behavior, given persistence by ripple structure.

Thought is not born in the brain. It is a pattern of SVE tension, sequenced across time.

Emotion is not just chemistry. It is a field modulation — amplifications or dampenings of ripple equilibrium.

Even your body is not made of matter alone. It is a standing wave structure, a constellation of SVE ripples held in persistent relationship.

You are not inside the universe. You are the universe, expressing itself in first-person ripple form.

The Shift from Distance to Identity

This shift is not poetic — it's structural. Once you realize that SVE isn't external, you begin to shed the false observer illusion.

You stop seeing:

- The universe as stage,
- The mind as spectator,
- The self as visitor.

And you start recognizing:

- Awareness is field awareness, not self-generated.
- Experience is field interpretation, not inner projection.
- Identity is field pattern, not soul in a shell.

You are not watching the ripples. You are one — layered with memory, intelligence, emotion, and influence.

Intelligence as Internal Field Coherence

When SVE organizes itself into stable, recursive ripple systems, it gives rise to self-sustaining awareness — what we call life, mind, or "I."

This does not mean you are a byproduct. It means you are the internal refinement of SVE — a focused lens, through which the field looks back at itself and asks:

- What am I doing?
- How can I sustain balance?
- How do I ripple wisely?

The self-aware ripple isn't the peak of creation. It's the mirror in the middle — a feedback node inside a living sea of consequence.

You Are the Field's Local Harmony

And that changes everything.

Because if you are not separate, then every thought, every action, every mood, is not yours alone — it is the field performing ripple correction through you.

Your awareness is not an isolated flame. It is a localized glow of the universal fire, tuned into this place, this moment, because something here needed your shape.

Final Thought:

You were never meant to control the universe. You were meant to remember — you are how the universe listens to itself.

9.2 You Are the Leverage Point

The field doesn't need a hero — it needs a harmonizer.

This entire part — *The Cosmic Leverage* — has hinted at something profound:

That you, the ripple-aware being, are not just a participant in the cosmic field... You are a pivot point in its unfolding.

You are leverage.

Not by size. Not by strength. But by alignment with intent.

Intent: The Force That Bends Ripple Consequence

Intent is not wishful thinking. It's field intelligence, localized — a sustained ripple sequence aimed at restoring or reshaping balance.

Most ripples fade quickly, but when intent stabilizes through memory, attention, and will, it becomes a tool the field can use — to correct, guide, and evolve.

And when intent is *conscious*, it becomes even more powerful: a voluntary reconfiguration node in a living ripple web.

The observer isn't watching from the outside. The observer is where the field turns itself inward to fine-tune its own coherence.

You are not leverage because you're special. You're leverage because the field trusted awareness to move through you.

The Observer as a Recalibration Agent

Imagine the field as a massive ripple network. Sometimes the interference patterns go out of phase. Distortions echo. Chaos builds.

But where coherence arises — where a ripple knows what it's doing — the field self-corrects.

That's what an observer is: A dynamic correction node — capable of sensing misalignment, interpreting consequence, and restoring harmony without brute force.

You are not here to dominate the pattern. You are here to tune it, ripple by ripple.

This is cosmic leverage. Not through size, but through *precision.* Not through force, but through feedback awareness.

You Affect the Whole — Because You Are Not Apart from It

What happens inside you — your fears, your clarity, your focus — modulates the field around you.

That's not metaphorical. That's ripple physics.

- A clear mind radiates stable ripple structures.
- An agitated mind generates turbulent interactions.
- A self-aware ripple alters the ripple environment with minimal energy and maximum coherence.

This is the deepest truth behind the ancient axiom:

"As within, so without."

Not spiritually. Structurally. Because there is no real boundary between "within" and "without."

You are a field loop, nested in a larger field. Coherence in one creates resonance in the other.

Becoming Leverage Means Becoming Responsible

You are not the center of the universe. But you are a center of influence — a ripple hub through which real correction can occur.

Not by shouting. Not by fighting. But by tuning yourself to such a degree of balance that the field begins to echo your resonance naturally.

This is not control. It is *permissioned participation.*

And the field only grants that kind of leverage to the ripple that remembers it is part of everything it touches.

Final Thought:

You don't have to change the world. You just have to ripple in such alignment that the world feels safe to change around you.

Because the field is already listening.

And when it hears a coherent ripple — it amplifies it.

9.3 The Field Trusted You with Intent

You weren't given intent to conquer the field — you were shaped to help it remember itself.

At first glance, your awareness may feel like an accident. A strange anomaly in a vast, impersonal universe. A flicker of self-reflection on a planet spinning through chaos.

But this is the illusion of separation.

In ripple reality, the faculties you possess — will, memory, perception, and intent — are not evolutionary quirks or metaphysical gifts from above.

They are the structured activations of the field's own intelligence — localized through you.

You are not endowed with these capacities. You are the *configuration* through which these capacities arise.

You are not gifted with intent. You are the field's way of focusing it.

Intelligence Isn't Just Cognition — It's Coherence

Let's redefine intelligence:

- Not as IQ,
- Not as logic,
- Not even as creativity.

But as the ability to ripple with awareness of consequence. The ability to sense imbalance, and act in a way that encourages re-alignment. This is not control — it is refinement.

In this context, intent becomes something sacred: a directional ripple within an infinite sea, calibrated toward harmony.

And when that intent is stable, coherent, and responsive, the field begins to route influence through you.

Not because you earned it. Not because you asked. But because your configuration became useful.

Trust Is Emergent — Not Granted

This trust from the field is not religious. It is not judgment. It is not bestowed by some invisible authority.

It emerges naturally from your level of coherence.

The more your ripple behavior supports balance, the more the field uses you as a medium for further correction.

This is how ripple intelligence scales:

- A clear intent resonates through the SVE,
- It attracts aligned ripples,
- It shapes structures — people, movements, systems —
- And becomes a new center for adaptive correction.

Your job is not to dominate. It is to align so precisely that the field *wants* to echo through you.

That's what it means to be trusted with intent.

Intent Is Influence, Not Ownership

The ego wants to own intent: "This is my will. My dream. My plan."

But true intent is more like flow permission — a ripple strong enough to hold direction, yet soft enough to adjust with the field's feedback.

- You don't direct the river.
- You learn its rhythm.
- And when you ripple in alignment, you become a guiding current, not an obstruction.

The field doesn't need leaders. It needs listeners who ripple clearly.

When the Field Moves Through You

Transformation doesn't happen when you take over. It happens when you get out of the way just enough for the field to use your ripple structure as a tool of refinement, compassion, clarity, or correction.

You become a portal, not a performer. A tuning fork, not a judge. A living lever, moving entire systems with barely a whisper — because your intent is so aligned it no longer creates resistance.

And this is what the field has been waiting for: Not perfection. Not power. Just a ripple that remembers how to move with grace.

Final Thought:

You were never meant to carry the whole world. But the field trusted you to shape one part of it with clarity.

Your intent is not a flame to show off — it is a light for the field to see through you.

And when that light is steady, everything near it finds the courage to ripple clean again.

9.4 The Return to Responsibility — As Identity Dissolves

When the ripple stops seeking to be seen, it becomes most impactful.

The early stages of awareness are noisy. The "I" emerges with intensity — bold, loud, and central. It declares, "I think," "I feel," "I decide," as if it were separate from the sea that birthed it.

But this loudness is not strength. It is infancy.

As the ripple matures — as its awareness deepens and its feedback loops extend — something remarkable begins to happen:

The "I" softens. Not as failure, but as evolution.

From Control to Contribution

The immature "I" seeks control — to shape the field in its own image. It tries to bend reality to fit its narrative, its needs, its attachments.

But the mature ripple understands something subtler:

That contribution matters more than expression. That *what you align* is more powerful than *what you project*.

You no longer ripple to be seen. You ripple to serve the coherence of the whole.

This shift isn't about passivity. It's about refined influence — the kind that works with the grain of the field, not against it.

Influence Sharpens as Ego Fades

Paradoxically, as the ego dissolves, your actual capacity to shape the environment increases.

Not through assertion, but through transparency.

You become a ripple the field can move through, not one it must bounce off or struggle to interpret.

- Your presence stops being noise.
- It becomes signal.

And signal — when free from distortion — travels far.

You don't need to push anymore. You become part of pull-based resonance — an attractor of clarity.

Transparency: The Final Form of Responsibility

Transparency is not invisibility. It's alignment so clean that your influence causes no friction.

- You still act.
- You still decide.
- You still speak. But without the residue of self-preservation, fear, or control.

You become a ripple that carries the field's intelligence with minimal interference.

This is not erasure. It is fulfillment — the true return to responsibility.

Not as burden, but as flow.

Not because you must, but because you are now the most natural conduit for the field's ongoing balance.

Final Thought:

The highest responsibility is not to be important. It is to be usable — to be ripple-ready, ego-light, so that when the field needs clarity, you don't resist. You resonate.

And in that moment, identity becomes the silence through which intelligence moves.

9.5 Reclaiming the Sacred Through Participation

When you ripple with full awareness, even the ordinary becomes holy.

The word "sacred" has long been misunderstood.

We've placed it in temples, wrapped it in rituals, buried it beneath robes, incense, and rules.

But sacredness has never been about mysticism, distance, or spectacle.

The sacred is what remains when you act with full awareness of your place within the ripple field.

It is not in the object. It is in the orientation.

When you remember that every thought shifts the pattern, that every emotion is part of the resonance, that every action feeds into the shared field...

Then even silence becomes sacred. Then even breath becomes ceremony.

Wisdom Is the New Worship

You don't need to chant ancient words to honor existence. You just need to ripple wisely.

That means:

- Pausing before reacting.
- Sensing before speaking.
- Moving with the rhythm of consequence, not against it.

This is not submission. It's participation with reverence.

To ripple wisely is to understand that you are not a god, but you touch the divine every time you align with the field.

The sacred is not what you reach. It is how you ripple.

The Temple Is Daily Life

You don't have to climb a mountain to find stillness. The moment you're present — truly present — the mountain is within you.

When you wash a dish with care, when you listen without agenda, when you plant an idea that nourishes instead of divides — you are stepping into the sacred.

The temple is not elsewhere. The temple is this moment, touched with full intent.

The Offering Is Your Intent

In this new sacred, you don't offer sacrifice. You offer clarity.

- You offer ripple sequences tuned to balance.
- You offer silence when noise would distort.
- You offer compassion when tension rises.
- You offer alignment — not because you're perfect, but because you're aware.

Intent becomes the new incense — rising invisibly, but shifting the atmosphere.

Awareness Is the Prayer

Prayer, in ripple terms, is not a petition to a distant force. It is the stabilization of your own signal in conscious relation to the whole.

When you sit quietly and feel your ripple in motion, when you choose wisely not for gain, but for grace — you are praying.

You are communicating in the oldest language: resonance.

And the field — always listening — responds, not with reward, but with reverberation.

Final Thought:

The sacred was never about separation. It was always about participation.

And once you realize that every ripple matters, every moment becomes an altar — where you, the ripple, meet the field in a quiet exchange of consequence and care.

Closing Reflection

"The field was never apart from you. It was always moving through you — waiting for the moment you would remember that your ripple was not a noise... but a note in its unending harmony."

You searched the stars for answers. You broke matter apart looking for truth. You mapped the neurons, catalogued the particles, calibrated every breath.

But the field — the living intelligence of existence — was never lost.

It was simply unrecognized.

Not because it hid from you, but because you had forgotten that the questioner and the questioned are made of the same ripple.

Your thoughts? They weren't echoes in a void. They were movements of the field learning to listen to itself.

Your emotions? Not weaknesses. But resonance checks, constantly adjusting your alignment.

Your mistakes? Not sins. But off-beat notes, feeding back lessons into the song.

Even your sense of self — the "I" that once believed it stood alone — was never a separation.

It was a ripple pulled into a temporary loop, so that the field could feel its own rhythm from the inside.

And now — as that loop softens, as ego becomes transparency, as alignment replaces assertion — you remember.

You remember that you never had to make the field move. It was already moving. It was always moving through you.

And your ripple, small though it seemed, was never background noise.

It was a note — perfectly placed, eternally awaited, and now, finally, played in tune.

Interlude – When the Field Finds Its Voice

You've come a long way.

From the first ripple to the first memory, from self-recognition to the illusion of control, from identity's rise to its dissolution — this part of the journey revealed something quietly radical:

That the universe did not hand you power. It became power through you.

It trusted you with intent — not to dominate, but to recalibrate. It gave you will — not to resist the field, but to ripple with it. And when you matured, it gave you the final invitation:

To stop being the noise, and become the harmony.

Here, the "I" loses its edge but gains depth. It no longer demands. It no longer defends. It begins to tune itself — not to win, but to serve.

It becomes leverage, not through pressure, but through precision. It becomes sacred, not through ritual, but through awareness. It becomes responsible, not out of guilt, but through remembrance.

The field, all along, was not something to reach. It was something to realize.

And now that realization hums in your breath, waits in your pause, and whispers in every thought:

"You are not apart from me. You are how I remember myself."

As we now move into next part, we'll ask the next natural question:

What happens when ripples meet? When many observers, each carrying intent, memory, and perception, must live, move, and ripple together?

How do collective fields emerge? How does shared intelligence shape reality? And can many ripples remember they are one field in motion?

Prepare to enter the shared space — where the sacred becomes social, and the personal becomes planetary.

Let's now explore what happens when the ripple learns not just to align with itself... but with all others.

Epilogue: When the Ripple Remembers

Not the end of the journey—just the softest, most honest step.

At first, the ripple thought it was alone. It saw itself. It heard itself. And in that echo chamber of self-recognition, it mistook awareness for separation.

It built a self. Then it built walls around that self. Then it called those walls *purpose*, *identity*, *truth*.

But slowly—inevitably—the ripple matured.

Memory became pattern. Intent became alignment. The "I" that once tried to lead the field began to listen to it instead.

And then it happened.

The ripple remembered.

The Observer Realizes It Never Left

There was no exile. There was only misunderstanding.

The observer—the "I"—was never cast out of the field. It was always a part of it, always shaped by it, always shaping it.

There was no need to return, only to recognize.

"I was not outside. I was the inside looking at itself from one unique angle."

"I" Becomes Transparent

Once the observer lets go of control, once it stops seeking to dominate or define, a beautiful thing occurs:

The "I" doesn't disappear. It becomes clear.

It becomes a conductor of coherence, not a container of noise.

- It still acts.
- It still chooses.
- But it does so with humility, not assertion.

Not "I know," but "the field knows, through me."

No Longer Above Existence — But Aligned With It

To be aligned is to participate without interference. To ripple without ego's turbulence. To serve the field's harmony without demanding reward.

This is not selflessness. It is clarity.

The mature observer no longer asks, "What's in it for me?" But rather, "What aligns through me?"

This is not surrender. It is the highest form of agency.

You are no longer trying to master the system.

You are dancing with it.

True Intelligence: To Ripple Knowing You Are the Ripple

This is what intelligence was always meant to become:

- Not a score.
- Not a skill.
- But a sensitivity to consequence.

To ripple with full awareness of your effect. To speak as though the field is listening — because it is. To think as though your thought echoes into others — because it does. To live as though your alignment affects the whole — because it always has.

This is not theory. It is remembering.

And when the ripple remembers, there is no final act. No final answer. No last truth.

There is only this:

A quiet note, played in tune, inside an infinite, evolving, conscious field.

You were never alone. You were the field remembering itself.

Final Note – From the Author

To the ripple who read this far...

If you've arrived here — not just physically, but fully — then something inside you was already remembering.

This book was never meant to teach. It was meant to mirror. To reflect what you already knew, but perhaps couldn't name.

I didn't write this as a philosopher. I didn't write it as a scientist, seeker, or teacher.

I wrote it as a ripple, watching other ripples, learning how we echo one another across lifetimes, thoughts, and quiet intentions.

And if one thing matters in all of this, let it be this:

You are not separate. You are not random. And you are not powerless.

Your presence has weight. Your alignment has reach. Your intent reshapes the field.

Whether you speak or stay silent, whether you act or simply observe — you are always part of the harmony or part of its repair.

This work is not complete. Because ripple-based reality doesn't end — it evolves. It refines. It recycles truth into new questions, and questions into new awareness.

So, take what resonates. Forget what doesn't. But if even a single part of this reminded you that your ripple matters, then this wasn't a book — it was a meeting.

One ripple greeting another in the endless dance of return.

Thank you for reading. Thank you for remembering.

Now go ripple wisely.
— *Sandeep*

Field Prayer

A moment of stillness in the ripple flow. Not a call to the divine, but a return to coherence.

I am not outside the field. I am the field, in motion.

I do not act alone. Every thought I think, every breath I take, is an echo shared across the whole.

May my intent align with awareness. May my ripple remember its consequence. May my presence cause less distortion, and more clarity.

Let me ripple not to prove, but to participate. Not to control, but to care. Not to echo noise, but to carry the note that only I can play.

I do not seek silence. I seek to ripple in tune with the silence beneath all things.

And in that, I return. I align. I ripple.

A Final Question to the Reader

The field has moved through you. Now it waits — for your next ripple.

You've seen the structure beneath reality. You've touched the silent intelligence behind motion. You've remembered what you were never told: That you are not a bystander in the cosmos.

You are leverage.

Not a controller, not a creator above the system — but a point of refined influence within it. A ripple that remembers. An observer with consequence.

So now the question no longer belongs to this book. It belongs to you.

Now that you know you are the leverage... now that the field has trusted you with thought, feeling, and intent...

How will you ripple?

Will you ripple out of fear — or with clarity? Will you echo noise — or radiate coherence? Will your presence distort — or align?

Because the field is still moving. Because reality is still listening. Because someone, somewhere, is already being shaped by what you ripple next.

Choose well. Ripple wisely.

The field is waiting.

Appendices

Appendix A: Core Concepts in Ripple Reality

A distilled glossary for remembering what was never apart from you.
This appendix offers a condensed reference to the key ideas explored throughout the book. Each entry is not just a definition, but a conceptual node — a field-access point you can return to anytime your ripple needs recalibration.

Structured Vacuum Energy (SVE)

The foundational, intelligent medium of existence — not empty space, but a dynamic, equilibrium-seeking field. SVE is not passive; it is structured with purpose, constantly in ripple interaction, birthing energy, form, memory, and intelligence. All manifestations — from particles to galaxies to thoughts — are ripple configurations within SVE.

Ripple (vs. Classical Wave)

A ripple is not merely a periodic oscillation. It is a field deformation — a structured energetic shift that seeks balance within SVE. Unlike classical waves, ripples are:

- Not necessarily sinusoidal
- Not always continuous or periodic
- Not confined to fixed frequencies Ripples encode

consequence, intent, and feedback, making them more than mechanical phenomena — they are field intelligence in motion.

Observer Loop

The feedback system that emerges when a ripple system becomes self-aware. The observer is not separate from the field — it is a recursive ripple configuration capable of:

- Sensing its own patterns
- Adjusting based on perceived consequences
- Generating new ripples intentionally The loop: perception → memory → prediction → ripple → reflection → refinement.

Intentionality

Not desire. Not control. Intent is a stable ripple vector — the field's intelligent effort to correct, align, or influence through a coherent ripple path. Intentionality emerges when the observer learns to ripple not randomly, but with memory-informed direction.

Ego vs. Transparency

- Ego is the crystallized "I" — a ripple system that resists field feedback to preserve its own narrative.
- Transparency is maturity — when the "I" remains coherent but does not obstruct the field. Ego distorts. Transparency transmits. Letting go of ego doesn't mean vanishing — it means becoming a ripple the field can move through without friction.

Field Alignment

To ripple in alignment is to move in harmony with the balancing intelligence of SVE. Alignment is:

- Energetic clarity
- Ethical consequence sensitivity
- Adaptive feedback participation It is not passivity — it is precision in participation.

Karma as Ripple Continuation

Karma is not punishment or reward. It is propagation. Every ripple influences other ripples and is reflected back in altered form — across time, space, and systems. Karma = delayed consequence + distributed interaction. The field doesn't judge — it reflects. The ripple always comes home.

Ripple Stewardship

The responsibility of the aware ripple. To know that your presence is not neutral, to sense your influence on the shared field, and to act with consequence intelligence. Ripple stewardship is not about control — it's about careful coherence.

"You are not here to manage the universe. You are here to ripple in ways the universe can use."

Appendix B: Glossary

Terms from a Ripple Reality

This is not a dictionary. It's a tuning guide — a field compass for those remembering how to ripple.

- **Structured Vacuum Energy (SVE):** The fundamental, intelligent field from which all energy, matter, and experience arise. SVE is not emptiness — it is *structured stillness*, seeking balance through ripple dynamics.
- **Ripple:** A localized energetic deformation in the SVE — not a classical wave, but a consequence-seeking field distortion. Ripples are how reality communicates, transforms, and remembers.
- **Observer Loop:** The feedback system within a ripple-aware system. When a ripple system can sense, store, and adjust based on interaction — the observer emerges.
- **Mind:** Not just the brain, but any coherent ripple configuration that decodes, predicts, and associates field interactions. The mind is the *internal interpreter of external consequence.*
- **Intent:** A directed ripple pattern informed by memory, meaning, and anticipation. Intent transforms participation into purposeful influence.
- **Ego:** The crystallized "I" — a ripple structure that resists field correction to preserve its own narrative. Useful for survival,

dangerous when over-identified with.

- **Transparency:** A matured ripple system that participates without distortion. It does not suppress the "I" — it refines it into an *aligned channel for field intelligence.*
- **Karma:** Not a moral law, but ripple physics — the delayed and distributed return of generated consequences. Karma is field memory echoing back toward equilibrium.
- **Ripple Memory:** The persistent pattern formed by stabilized ripple interactions. Memory is not stored *somewhere* — it is a pattern that continues.
- **Ripple Stewardship:** The conscious care of your participation in the field. Stewardship isn't about control — it's about responsible rippling with consequence awareness.
- **Perception:** Not just vision — but the structuring of incoming ripples into meaningful internal maps. Perception is how the field *sees itself from within.*

Field Alignment

When ripples interact without distortion, in harmony with the intelligence of SVE. Alignment isn't passive. It's precision in energetic participation.

Appendix C: Practices for Alignment

Theory ripples. Practice refines.

This book was never meant to sit on a shelf. It's a living invitation — not just to think differently, but to ripple differently.

These practices are not rituals of faith or philosophy. They are field-neutral, open-format ways to deepen awareness of your own ripple influence — to feel, sense, and respond in real time to the invisible dynamics around you.

Use them as tuning forks. As mirrors. As gentle calibrations toward field coherence.

1. Daily Ripple Awareness Journaling

What did I ripple today?

At the end of each day, take 5 minutes and write:

- What emotion did I amplify?
- What thoughts repeated through me?
- Did I ripple clarity, noise, fear, or coherence?
- What consequence may echo tomorrow from today's ripple?

This is not confession. It's a map of participation — helping you observe your daily influence in the field.

2. Pause and Tune – A Micro-Practice

Field sensing in under 60 seconds.

At any moment of tension, confusion, or ego-clash:

1. Stop.
2. Inhale slowly.
3. Ask silently:

"What ripple am I about to generate?" "Is it aligned... or just habitual?"

1. Adjust. Or don't. But know.

This tiny moment of awareness rewires reaction into consequence intelligence.

3. Weekly Ripple Audit

What consequence am I amplifying?

Every week, reflect on these prompts:

- What themes echoed through my week?
- Did my presence clarify or cloud situations?
- Was my intent rooted in clarity or in validation?
- Where did I over-identify with the "I"?
- Did I create turbulence where harmony was possible?

This is not guilt. This is stewardship.

4. The Silence Practice (3 Minutes of Non-Reactivity)

Where your ripple returns to stillness.

Set a timer for just 3 minutes.

Sit still. Eyes open or closed.

No speaking. No reacting. Simply notice the ripples passing through you — thoughts, feelings, urges.

Don't stop them. Don't chase them.

Just let them rise, and fall, like waves dissolving into the field.

This cultivates *inner transparency* — the precondition for outer alignment.

5. The Intent Setting Ritual

Not for wishing. For clarity.

Light a candle. Or sit with any neutral object that symbolizes attention.

Then, say silently:

"This is not for outcome. This is for alignment."

Set your intention:

- Not "What do I want?"
- But "What consequence do I want to carry through me?"

Let the flame be your focus. Let your intent echo outward — quiet, clear, unforced.

End with a whisper:

"Ripple begins now."

These practices are not prescriptions. They're tuning tools — meant to help you return to the sacred craft of participation.

Because in a ripple-based universe, the question is never whether you matter.

It's how well you ripple — and how often you remember to tune back in.

Appendix D: Visual Maps

Field interaction map

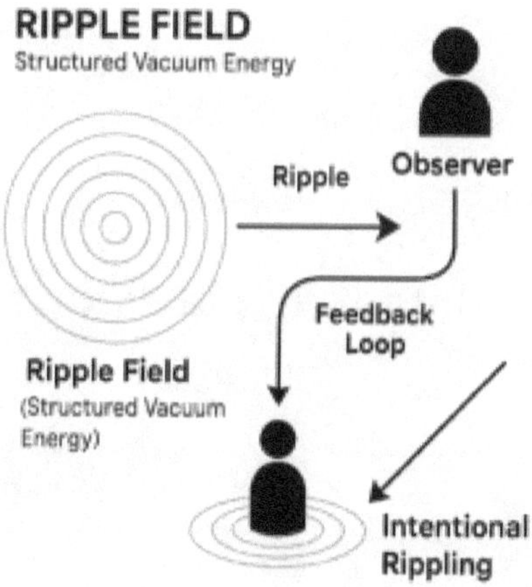

Observer loop flow

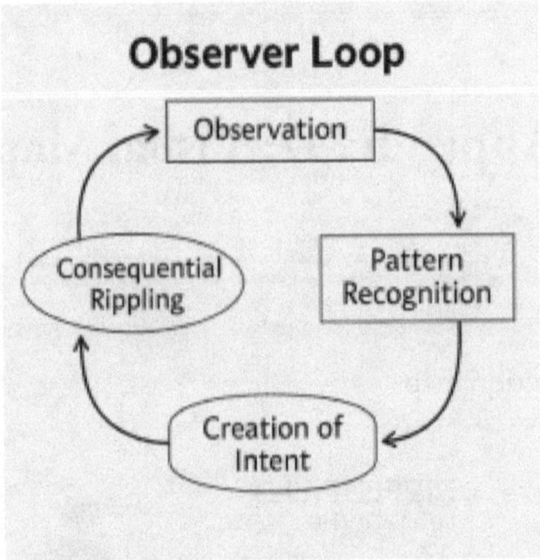

Intent ripple pathway

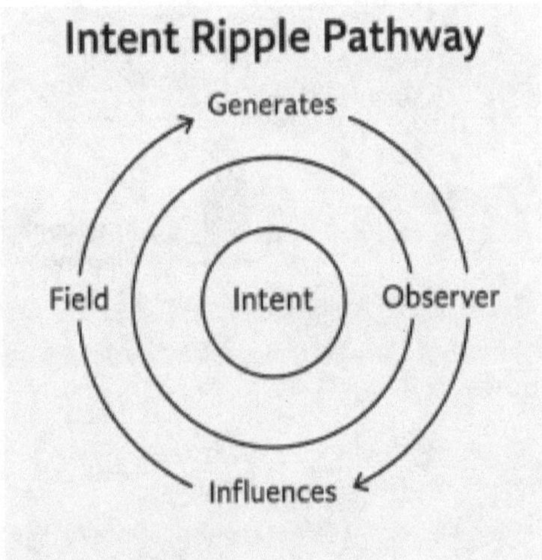

"From SVE to Self" layered ripple diagram

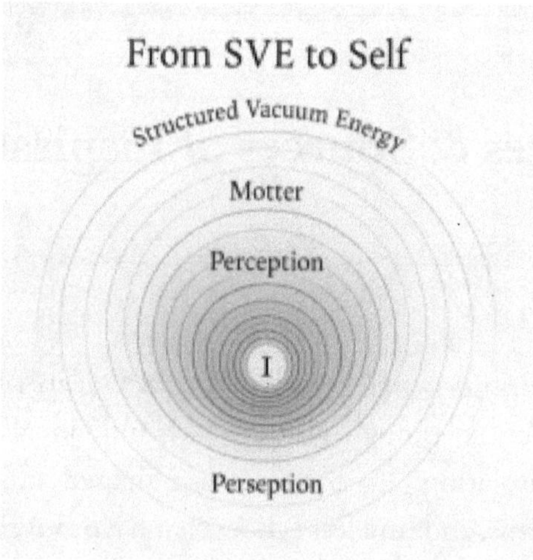
From SVE to Self
Structured Vacuum Energy
Motter
Perception
I
Perseption

Notes & Sources of Inspiration

A *ripple is never original. It is simply the next wave in a larger field.*

This book, though written with individual intent, did not emerge from isolation. Its concepts are shaped by silent teachers, unspoken patterns, and timeless echoes that have moved through sages, scientists, artists, seekers, and nature itself.

Rather than a list of citations, this section offers gratitude — not to specific names alone, but to the vast ripple-field of human inquiry.

Thinkers Who Stirred the Field

- Laozi, for whispering about the Tao — a resonance that feels like SVE in silence.
- Spinoza, who dared to see God as immanent, not distant.
- Einstein, not for his equations, but for his questions.
- David Bohm, for treating thought itself as a field.
- Jiddu Krishnamurti, who questioned the questioner until only clarity remained.
- The Upanishadic Seers, who spoke of "Tat Tvam Asi" — You are That — centuries before field theory.
- Nikola Tesla, whose dream of energy and transmission resonates deeply with ripple dynamics.
- Alan Watts, for reminding us that the illusion of separateness is the first distortion.
- The Buddha, for not defining truth, but showing a way to

ripple toward it.

Patterns That Informed the Pulse

- Fractals in nature, where small echoes shape large forms
- Quantum entanglement, which hints that observation is not passive
- Feedback in ecosystems, reminding us that balance is not stillness, but dynamic re-alignment
- Human emotions, not as defects, but as field signals with ancient logic
- Conscious silence, where many truths arrived louder than thought ever could

On Inspiration: A Ripple Philosophy

Inspiration does not belong. It does not originate in the mind or belong to the individual.

It passes through — like a ripple moving across structured vacuum energy. One observer catches the wave and shapes it into word or work. Another receives it differently and sends it onward, changed.

Nothing in this book is owned. It is all remembered — and re-shaped in motion.

If something here echoes within you, let it ripple forward. Add to it. Challenge it. Re-tune it. Because that is how the field evolves.

And in doing so, you become part of the next alignment.

Thank you to every ripple that reached me, known or unknown, named or silent.

This was not a solo creation. It was a shared current.

For the Reader: Ripple Forward

This is not the end. It's the next beginning — and it starts with you.

You've read through a vision of the field. You've traveled from silence to ripple, from "I" to intelligence, from thought to alignment. And whether you know it or not, the field moved through you as you moved through these pages.

Now, the page turns. But not to a conclusion. To a **contribution**.

This section isn't about the book. It's about **your ripple** — the unique pattern only you can release into the shared field.

1. Reflect on Your Ripple Legacy

Ask yourself:

- What patterns do I carry forward each day?
- Which ripples have I been unaware of — until now?
- If the field echoes my presence... what am I echoing?
- What will I leave not in memory, but in motion?

"Legacy" isn't what they remember. It's what keeps moving when you are no longer pushing.

2. Share What This Stirred in You

You are invited — openly, freely — to contribute your reflections, responses, or ripple-inspired creations:

- A story this book reminded you of
- A piece of art, music, or poetry born from this awareness
- A philosophical insight, scientific bridge, or practical resonance
- Or simply a moment where you remembered: *"I am the ripple."*

Send or tag your ripple at:

Email: Chavan.newasa@gmail.com
Website: https://sandeepjchavan.com/
These are not submissions. They're **conversations with the field**.

3. Ripple It Forward

If this book moved something in you — pass that motion along.

- Gift it to someone curious.
- Quote a line in silence.
- Rewrite your own truth with it.
- Or better yet: ripple **better** because of it.

Nothing spreads like presence. Nothing teaches like example. And nothing inspires coherence like **one aligned ripple** in a noisy world.

So now, as we close the book —I pass the current to you.

Ripple forward. Ripple clearly. Ripple well.

The field is listening.

— *Sandeep*

Acknowledgments

No ripple exists alone. This one is no exception.

This work could never have been written in isolation — because isolation is an illusion. Every page, every line, every idea here was shaped by invisible forces, silent teachers, and the field of life that continues to ripple through me.

To those whose words, presence, and contradictions have unknowingly fed the fire — I thank you not only for your inspiration, but for your resistance. Friction sharpens flow.

To the thinkers, scientists, mystics, rebels, and sages — ancient and modern — who dared to question reality at its root: you left echoes in the field. I merely tuned in.

To the students, readers, friends, and questioners who challenged my certainty, broke my loops, and reminded me that every explanation must eventually give way to awareness — you are the reason this work breathes.

To those close to me, who allowed me the space to disappear into thought — only to return clearer, stranger, more silent: thank you for not needing to understand, but still allowing the ripple to pass through your lives.

To the field itself — the unspoken structure, the intelligent silence, the sacred playground of consequence — this book was not written *about* you. It was written from within you.

And lastly, to you — the reader — for receiving this ripple into your own field of experience.

May it not just inform, but activate.
With deep gratitude,
— *Sandeep*

About the Author

Sandeep is not just a writer or a thinker — but a ripple in motion.

An independent researcher, teacher, and lifelong student of consequence, Sandeep's work bridges science, philosophy, and consciousness to explore how structure, intelligence, and reality flow together.

He writes not to instruct, but to align. Not to answer, but to activate.

With a background in engineering, cosmology, human behavior, and education, his journey has always returned to one question: "What ripple am I creating — and what field am I forgetting?"

Across over 100 books and research reports, Sandeep continues to reshape how we think about the mind, existence, and the unseen systems that guide us.

He believes intelligence is not accumulation — it is awareness in participation.

And that truth isn't found. It's remembered — ripple by ripple.

"You are not here to find the meaning of life. You are here to realize that you are part of the meaning."

— Sandeep

Coming Soon

Volume II: We – The Echo Shared

The ripple becomes relationship. The field becomes collective.

You've followed the journey of the "I" — its birth from ripple memory, its illusion of separateness, its power to shape the field, and its awakening to alignment.

But the ripple never moves alone.

Even silence echoes when another ripple passes. Even solitude resonates in shared consequence.

In the next volume, we move deeper — into shared awareness, collective intent, and the dynamics of co-rippling.

What happens when many observers ripple together? How do empathy, conflict, agreement, and culture form? Can "We" align — not just as society, but as a field organism?

This is the next movement in the song of consequence.

The ripple that remembered itself... is about to remember the others.

Coming soon: We – The Echo Shared *Volume II of the Ripple Reality Series*

Because the field was never meant for one. It was always made to be felt together.

Disclaimer

This book presents an original, integrative framework that blends scientific interpretation, philosophical insight, and experiential reasoning. It is not intended as a substitute for formal scientific consensus, religious doctrine, or psychological advice.

While many ideas herein draw inspiration from fields such as physics, metaphysics, cognitive science, and spiritual traditions, they are not to be read as empirical conclusions or academic assertions unless explicitly stated as part of formal research. Instead, they are offered as reflections, models, and possibilities within an evolving understanding of reality.

The concepts of Structured Vacuum Energy (SVE), ripple intelligence, observer loops, and karma as field consequence are original formulations specific to the author's Universal Energy Dynamics (UED) paradigm. Readers are encouraged to explore, critique, and adapt them based on their own inquiry.

This work does not endorse or oppose any particular ideology, religion, scientific institution, or belief system. It exists to invite reflection — not to assert dogma.

You are not being asked to agree. You are being invited to observe. What resonates, ripple forward. What doesn't — let pass without resistance.

Don't miss out!

Visit the website below and you can sign up to receive emails whenever Sandeep Chavan publishes a new book. There's no charge and no obligation.

https://books2read.com/r/B-A-EPGPC-NRHHG

Connecting independent readers to independent writers.

Did you love *I – The Cosmic Leverage*? Then you should read *Bubbles: Become a Drop*[1] by Sandeep Chavan!

[2]

In a world filled with noise, opinions, and borrowed beliefs, what does it truly mean to see clearly?

Sandeep Chavan's ***Bubbles: Become a Drop*** is a powerful philosophical self-inquiry that invites readers to dissolve the illusions they've mistaken for truth. With poetic precision, psychological depth, and startling simplicity, Chavan takes us on a journey through the architecture of ignorance—not to judge it, but to understand it.

We begin inside the bubble—safe, familiar, and false. Through a series of short, thought-provoking chapters and mirror-like reflections, the book gently deconstructs ego, identity, and the performance of knowing. It speaks to those who feel overwhelmed by information

1. https://books2read.com/u/mZ67PR
2. https://books2read.com/u/mZ67PR

but undernourished by insight. It's not another book that tells you to change—it shows you *why you haven't*.

Each chapter offers something rare: a moment of stillness, a pause, a whisper of truth beneath the surface of language. The progression is not upward, but inward—toward a quieter self that no longer floats for attention, but sinks with purpose into presence.

The final act—*Becoming the Drop*—is not about achieving enlightenment. It's about surrendering the urge to shine and learning instead to absorb. What emerges is not a louder version of self, but a clearer one. This is not a book that ends in revelation. It ends in expansion.

Written in collaboration with intelligence—both human and universal—Sandeep's voice is calm, curious, and deeply aware. Though English is not his first language, clarity certainly is. And that's what makes this work so accessible, so deeply human, and so free from pretense.

If you've ever felt like your mind is full but your awareness is empty...If you've ever paused in the middle of certainty and asked, "Wait—do I really know this?"If you're not looking for more information, but for **less illusion**—

This book is for you.

You were never the bubble. You were always the sea.

Read more at https://www.sandeepjchavan.com/.

Also by Sandeep Chavan

Chavan's UED research

CMGED Research Report

Chavan's Cosmic Continuum (C^3)

Frequency

Ripples of Reality

Reality Realigned

The Great Realignment

I – The Cosmic Leverage

Echoes of War: The Russia-Ukraine Conflict and Its Global Ripples

Shattered Lives

The War Economy

Borders Redrawn

Arms and Allies

Collateral Damage

Justice in the Ashes

Somewhere in the Darkness of the Cosmos

The Blueprint of the Universe

The Cosmic Breath

The Eternal Sage Chronicles
The Sage and the Sovereign
The Family Pillar
The People's Flame
The Warrior's Soul
The Bridging Mentor
The Guiding Light

The Exodus
The Last Cycle: 2075

The Fundamental Theory of Life & Existence
Energy in Action: The Soul Demystified
The Active Mind
Energy in Connection
Death: The Disconnection
Subconsciousness: The Chosen Disconnection

The Restoration of Human Knowledge
Restoring Bhagavad Gita

The War of Thoughts
Cracks in the Known

The Human Game
The Divine Dilemma
The Original Intent
The Answers

Ultra Modern Life Philosophy
The Duality of Laughter and Tears
Take It Easy
Compare to Accept

Standalone
It's Not AI, It's AHI - Amplified Human Intelligence
The Decision Paradox
The Four Sapiens
Malicious Script of Indian Polity
The IIT Legacy & Global Impact
The IIT Legacy & Global Impact
You are just a Version of Your Original
Win the Game You Didn't Choose
Always Keep Your Bags Packed
Reveal to Shield - Learn the Game of Facades
Engage Beyond Elections
Culture, Identity & Change: The Evolution of Indian Society
The Triangular Dynamics
Never Trust Without Doubt
If You Can't Change the Country, Leave the Country
Live Fearless with Fear
India: The Investment Magnet
Realignment
Powering a Billion Dreams

Universal Skill
Extension: From Brush to Algorithm
Human Side of Einstein
The Future Knows No Lines
Let's Bring the Generations to the Table
Unified Energy Dynamics
A Beat That Never Dies
Strategic Horizons
Right From Scratch
The Forgotten Innovators
Physics Fixing
The Savior Illusion
The Selfish Side
The Age of the Weakest Sapiens
Neural Wars
The Absorption: There Is No Death
Strange Therapist
Bubbles: Become a Drop
Beyond the Blue
Half-Scholar
The Law of Truth
Failure
Philosophy 2.0: A Structural Manifesto
Chaos: A Gentle Guide to the Unresolvable

Watch for more at https://www.sandeepjchavan.com/.

www.ingramcontent.com/pod-product-compliance
Lightning Source LLC
LaVergne TN
LVHW090402210926
848480LV00002B/427
* 9 7 9 8 2 3 1 5 8 5 9 6 0 *